T0332602

DE MOTU AND THE ANALYST

The New Synthese Historical Library
Texts and Studies in the History of Philosophy

VOLUME 41

The titles published in this series are listed at the end of this volume.

GEORGE BERKELEY

DE MOTU
AND
THE ANALYST

A Modern Edition, with
Introductions and Commentary

Edited and translated by

DOUGLAS M. JESSEPH
North Carolina State University, Raleigh, U.S.A.

KLUWER ACADEMIC PUBLISHERS
DORDRECHT / BOSTON / LONDON

Library of Congress Cataloging-in-Publication Data

Berkeley, George, 1685-1753.
    [De motu]
    De motu, and, the analyst / by George Berkeley ; a modern edition
with introductions and commentary by Douglas M. Josseph.
        p.   cm. -- (The New synthese historical library ; v. 41)
    Includes index.
    ISBN 0-7923-1520-0 (alk. paper)
    1. Science--Philosophy. 2. Mathematics--Philosophy.   I. Jesseph,
Douglas M.   II. Berkeley, George, 1685-1753. Analyst.   1992.
III. Title.   IV. Series.
Q175.B4674 1992
501--dc20   91-37667

ISBN 0-7923-1520-0

Published by Kluwer Academic Publishers,
P.O. Box 17, 3300 AA Dordrecht, The Netherlands.

Kluwer Academic Publishers incorporates
the publishing programmes of
D. Reidel, Martinus Nijhoff, Dr W. Junk and MTP Press.

Sold and distributed in the U.S.A. and Canada
by Kluwer Academic Publishers,
101 Philip Drive, Norwell, MA 02061, U.S.A.

In all other countries, sold and distributed
by Kluwer Academic Publishers Group,
P.O. Box 322, 3300 AH Dordrecht, The Netherlands.

Printed on acid-free paper

Printed in the Netherlands

For Doreen

Table of Contents

THE ANALYST

Preface

Berkeley's philosophy has been much studied and discussed over the years, and a growing number of scholars have come to the realization that scientific and mathematical writings are an essential part of his philosophical enterprise. The aim of this volume is to present Berkeley's two most important scientific texts in a form which meets contemporary standards of scholarship while rendering them accessible to the modern reader. Although editions of both are contained in the fourth volume of the *Works*, these lack adequate introductions and do not provide complete and corrected texts. The present edition contains a complete and critically established text of both *De Motu* and *The Analyst*, in addition to a new translation of *De Motu*. The introductions and notes are designed to provide the background necessary for a full understanding of Berkeley's account of science and mathematics.

Although these two texts are very different, they are united by a shared a concern with the work of Newton and Leibniz. Berkeley's *De Motu* deals extensively with Newton's *Principia* and Leibniz's *Specimen Dynamicum*, while *The Analyst* critiques both Leibnizian and Newtonian mathematics. Berkeley is commonly thought of as a successor to Locke or Malebranche, but as these works show he is also a successor to Newton and Leibniz.

Another obvious similarity between these two texts is their predominantly critical stance: both are essentially critiques of the scientific and mathematical theories of the early eighteenth century, with comparatively little emphasis on a positive account of Berkeley's views. A full account of Berkeley's science and mathematics is a task for another time and place, but any proper understanding of his philosophical conception of these subjects must begin with *De Motu* and *The Analyst*.

I am happy to acknowledge the support of the National Endowment for the Humanities for two grants: a travel grant in 1989 which helped send me to the British Library, and a summer fellowship in 1991. The British Library, the Special Collections Division of the Regenstein Library at the University of Chicago, and the Bodleian Library all deserve thanks for providing access to primary sources. The Clapp Library of Wellesley College provided an excellent microfilm of *The Analyst*, while the Special Collections Division of Regenstein allowed me to use their copy of *De Motu*. With their permission I have reproduced the title pages from both works. Thanks are due to the proprietors and regular patrons of the Museum Tavern in Bloomsbury, most especially to John Henderson and Brian Rooney. Good beer and good conversation in the evenings made two research trips to the British Library more pleasant than I had reason to expect. Dan Garber and Ken Winkler provided advice, assistance, and encouragement which I greatfully acknowledge; correspondence with Lisa Downing was helpful in sorting out my thoughts on *De Motu*. By far my greatest debt is to my wife, Doreen, whose patience, good nature, and assistance were invaluable.

Abbreviations

AG Leibniz, G. W. 1989. *G. W. Leibniz: Philosophical Essays*. Ed. and trans. Roger Ariew and Daniel Garber. Indianapolis, IN and Cambridge, MA: Hackett Publishing Company.

AT Descartes, René. [1897-1909] 1964-1976. *Oeuvres de Descartes*. Ed. C. Adam and P. Tannery. Paris: Vrin/C.N.R.S. 12 vols.

CSM Descartes, René. 1984. *The Philosophical Writings of Descartes*. Ed. John Cottingham, Robert Stoothoff, and Dugald Murdoch. 2 vols. Cambridge: Cambidge University Press.

GM Leibniz, G. W. [1849-55] 1962. *G. W. Leibniz: Mathematische Schriften*. Ed. C. I. Gerhardt. 7 vols. Hildesheim: Olms.

apers Newton, Isaac. 1967-1981. *The Mathematical Papers of Isaac Newton*. Ed. D. T. Whiteside and M. A. Hoskins. 8 vols. Cambridge: Cambridge University Press.

cipia Newton, Isaac. [1729] 1934. *Sir Isaac Newton's Mathematical Principles of Natural Philosophy and his System of the World*. 2 vols. Trans. Andrew Motte; rev. Florian Cajori. Berkeley and Los Angeles: University of California Press.

orks Berkeley, George. 1948-57. *The Works of George Berkeley, Bishop of Cloyne*. Ed. A. A. Luce and T. E. Jessop. 9 vols. (Edinburgh and London: Thomas Nelson).

Quotations from Aristotle come from *The Complete Works of Aristotle: The Revised Oxford Translation*. Ed. Jonathan Barnes. 2 vols. Princeton: NJ, Princeton University Press, 1984. Citations include the Bekker page numbers for the Greek followed by volume and page numbers for the Revised Oxford Translation.

De Motu

Editor's Introduction

In 1720 the Royal Academy of Sciences in Paris announced a prize essay competition on two topics, the first being the nature, principle, and cause of the communication of motion. Berkeley was in Europe at the time, acting as tutor to George Ashe, son of the Bishop of Clogher. Joseph Stock's biography reports that "On his way homeward [Berkeley] drew up at Lyons a curious tract *De motu*, which he sent to the royal academy of sciences at Paris, the subject being proposed by that assembly."(Stock [1776] 1989, 19) This was the first prize offered by the Academy, and it was awarded to Jean-Pierre Crousaz, professor at Lausanne. The decision seems to have been somewhat controversial, because the indifferent effort of Crousaz won out against competiton from several more competent entries, including one from Jean Bernoulli.[1] No record of Berkeley's entry survives in the archives of the Academy of Sciences, and we cannot be certain whether it was ever submitted.[2] Despite the apparent failure of the essay to win favor in Paris, Berkeley thought highly enough of it to publish *De Motu* in London in 1721, but again it failed to arouse interest in scientific and philosophical circles.

---

[1] Bertrand ([1869] 1969, 184-5) reports that "Les concurrents devaient traiter du principe, de la nature et de la communication du mouvement. Jean Bernoulli concourut; l'Académie, sans comprendre la portée de son excellent mémoire, couronna le discours superficiel et insignificant d'un M. de Crousas. L'injustice était flagrante, ou plutôt la méprise."

[2] Claudine Pouret, *documentaliste* in the archives of the Académie des sciences, reports in a letter of 24 January, 1990 "Je n'ai trouvé aucune trace de ce mémoire." The earliest manuscripts of prize essay submissions date from 1760 and the first register of submissions dates from 1745, so it is impossible to tell whether Berkeley actually submitted his work. Crousaz's piece was printed in the first volume of the collection *Recueil des pièces qui ont remporté les prix de l'Académie Royale des Sciences*. An expanded French version appeared as (Crousaz, 1728).

In accordance with the topic of the prize essay competition, Berkeley's *De Motu* proposes to treat the "principle and nature of motion," as well as the "cause of the communication of motions." The seventy-two sections of the text fall into three corresponding divisions: §§1-42 on the principle of motion, §§43-66 on the nature of motion, and §§67-72 on the cause of the communication of motion. These topics were familiar territory in philosophical discussion from the seventeenth and early eighteenth centuries, when "natural philosophy" saw the rise of a mechanistic paradigm for the explanation of physical phenomena. Although the scientific developments of the seventeenth and eighteenth centuries are far too varied and complex to be adequately summarized here, the broad outlines of some problems in the theory of motion must be sketched in order to set the background to Berkeley's *De Motu*.[3] As we will see, Berkeley's *De Motu* addresses a number of controversial issues and an understanding of these controversies is essential for understanding Berkeley's aims and his program for physics. I will begin with a brief account of central themes in the rise of the "mechanical philosophy" and will then deal with the specific issues addressed in *De Motu*. Later, I will turn to the question of the place of this work in Berkeley's philosophy.

1. MOTION AND THE MECHANICAL PHILOSOPHY

The greatest scientific treatise of the seventeenth century is undoubtedly Newton's *Mathematical Principles of Natural Philosophy* of 1687, which presents an entire system of the world governed by basic laws of motion. These laws unify terrestial and celestial phenomena by reducing all motions to the same basic principles, so that the explanation of the moon's orbit is (in principle) no different than the account of projectile motion near the surface of the Earth. It is no accident that Newton's treatise should begin with a statement of fundamental definitions and laws of motion: the basic problem for any physical theory

---

[3] See Dijksterhuis (1961), Hall ([1962] 1968), Koyré (1957), Westfall ([1971] 1978), and Westfall (1971), for general studies of the history of science in the period, and particularly problems in mechanics.

is to explain how and why objects in the world move in their characteristic manners. Of course, Newton's work fits into a pattern of development in the seventeenth century natural philosophy which has been variously called the "mechanization of the world picture" or the "mathematization of nature." In any case, Newton and his predecessors (most notably Galileo and Descartes) saw themselves as working on an entirely new approach to physics which would emphasize mechanical processes and their mathematical formulation rather than the discredited methodology of the scholastics. Berkeley is sympathetic to much of the mechanical philosophy, although he does critique it in many points of detail. One important feature of his relationship to the mechanical philosophy should be stressed at the outset: when Berkeley disagrees with Newton and other "moderns," his usual accusation is that the proponents of the new physics have introduced concepts and principles as obscure and confused as those of their scholastic predecessors. To make sense of Berkeley's critique of the mechanist program, we must therefore begin with a brief presentation of Aristotelian and Scholastic theories of motion, after which we can proceed to an outline of the principal components of the new science of motion in the seventeenth century.

1.1 ARISTOTELIAN AND SCHOLASTIC BACKGROUND

Motion as we understand it today is characterized by Aristotle as *local motion* or change of place. Aristotle's general account of change distinguishes change of place from other kinds of change, such as change of substance (as when a burning candle becomes smoke and flame), change of quality (as when an object changes color), or change of quantity (through growth or shrinkage). All of these are characterized as motions, but local motion is the most important kind and will be the object of our concern here.[4] The starting point for the Aristotelian doctine of local motion involves the doctrine of four elements and the associated theory of natural place. According to the four-element doctrine earth, water, air, and fire are the pure elements out of which all

---

[4] See Clagett (1959) and Grant ([1971] 1977) for an introduction to the medieval science of motion.

material substances are composed. Moreover, each pure element has a natural place: earth at the immobile geometric center of the universe, water above the earth, with air and fire naturally belonging in successive layers above water. These natural elements are contained within the lunar sphere, which marks the boundary between the incorruptible heavens (composed of a separate fifth element) and the sub-lunar world where things are in a constant state of change.

If the elements appeared only in their pure form each would naturally seek its own place, and the sub-lunar world would be reduced to concentric spheres of earth, water, air, and fire. But natural substances are composites, so that any given material substance contains a blend of several elements. The composition of any given material substance is dicated by its substantial form – the organizing principle from which the characteristic properties of the substance derive. The substantial form of a stone, for example, presumably dictates that it is predominantly composed of earth, with an admixture of other elements. Then, the question "Why does the stone fall downward when left unsupported?" has an easy answer: being largely earth, the substance seeks its natural place near the center of the universe and hence travels downward through the air. Because the world of material substances has many different kinds of things, there is a tremendous variety of natural motions present in the world.

The doctrine of natural place was combined with several other Aristotelian principles of motion, most notably the doctrine that everything that moves is moved by something else. This principle rules out a universe of perpetually self-moving objects and makes the mover or "motive power" distinguishable from the body moved. Further, the motive power may be contained within the moving body or act upon it by direct contact, but "action at a distance" is ruled out as impossible. Animals appear to be self moving, and might be thought to violate the principle that everything in motion is moved by something else; but the animal body is moved by the soul and in such cases we can distinguish the principle of motion from the body moved, even though the animal contains its motive principle.

Inanimate bodies present a somewhat different case. In the first place, the motions of inanimate bodies can be distinguished into the categories "natural" and "violent." Natural motion is that which a body manifests when it seeks its natural place: the upward movement of fire and the downward movement of a stone are paradigmatic natural motions. In the sub-lunar sphere rectilinear motion toward an object's natural place falls under the category of natural motion, while the circular motion of celestial objects is another kind of natural motion. But not all motion in the sub-lunar sphere is natural. A stone thrown from a sling does not move naturally, since its motion is initially away from the center of the Earth. All such violent motions seem to satisfy the Aristotelian dictum that everything in motion is moved by something else, because a stone does not spontaneously take off toward the sky and then return. But once the stone is released from the sling it is no longer in contact with the body which acts as the motive power, and we have an apparent case of action at a distance. Thus, it seems that some other motive power must be found to explain the continued (if gradually diminished and ultimately reversed) motion of the stone away from its natural place. Aristotle explained the continued motion of a projectile in terms of the action of the medium through which it travels: the initial motion imparted by the sling moves the stone, which in turn moves the air in front of the stone; but since he denied the possibility of a vaccuum, Aristotle reasoned that the space behind the moving stone is filled in by other bodies which squeeze the stone forward. Eventually, the natural resistance of the medium overcomes the initial motive force implanted by the violent motion, and the stone returns to its natural place.

The natural motions of inanimate bodies seem less in need of explanation, but they do pose some conceptual problems. In analogy with animate bodies whose principle of motion is the soul and is contained within the moving body, the principle of natural motions could be the substantial form of the body moved. Aristotle, however, identified the primary cause of natural motion with the particular agent or cause which had produced the body. Thus, the natural process which produces a stone endows it with its essential properties, including its tendency to move toward the center of the Earth. The proximate cause of the motion is the substantial form of the body, but the genuine moving power is the *generans* which first produced the inanimate body. Another problem for the theory of natural motion is the acceleration of a decending body. As heavy bodies approach the center of the Earth, their velocity increases; but distance from the Earth cannot be a cause of the accelerated motion, for that suggests that the Earth acts at a distance on the moving body. There were many different theories advanced to explain the cause of acceleration, and much of the background to modern physics can best be characterized as the search for a plausible account of naturally accelerated motion.

Medieval studies of motion took place within the context of Aristotle's theory, although there was hardly unanimity of opinion. In particular, Aristotle's accounts of projectile motion and falling bodies were criticized by various thinkers and modified in several ways. The most significant scholastic development for our purposes is the *impetus* theory of projectile motion, which is most closely tied to the work of Jean Buridan.[5] On Buridan's account impetus is imparted to a moving body when violent motion begins; this acts as a motive force to propel the body forward and is proportional to the quantity of matter in the body and the speed with which it is moved. The impetus acts as a "nonpermanent form" impressed on the moving body and is diminished by resistence, either from the surrounding medium or from so-called "internal resistence" on the part of the body itself. Internal resistence arises from the composite nature of bodies: being a mixture of different elements, each with an associated natural place, a body will have different tendencies upward and downward, and so manifest an internal resistence to different motions. Whatever the source of the resistence, Buridan reasoned that the quantity of impetus was fixed once the body had been set into violent motion and would eventually be overcome by a contrary tendency.

Buridan also applied the impetus theory to the problem of explaining the acceleration of descending bodies. He reasoned that the weight of the body is the primary cause of a body's fall, but because weight is constant this seems to leave the cause of acceleration unaccounted for. On Buridan's analysis, however, the heaviness of a body not only initiates its descent, but also acts continuously as it descends. Thus, at any point along path of descent, a body will have accumulated successive increments of impetus or "accidental heaviness" which will in turn generate successive increments of velocity. The acceleration of a falling

---

[5] See Clagett (1959, 505-540) for details of Buridan's work.

body is then treated as a consequence of the continuous action of the initial cause of its descent.

Berkeley apparently had no detailed knowledge of Aristotelian and scholastic theories of motion, and we should not expect to find a careful criticism of the doctrine of natural place or the impetus theory of motion in *De Motu*. Nevertheless, he does make occasional references to Aristotle and condemns "the obscure subtlety of the scholastics," in §40. The root of Berkeley's hostility to such accounts of motion is that they introduce meaningless terms in the attempt to explain motion, so that talk of substantial forms or impetus merely confuses the issue. In this respect Berkeley's views are hardly novel, but he brings the same complaint against more recent accounts of motion, and it is to a summary of these that we must now turn.

1.2 GALILEO, DESCARTES, AND THE INERTIAL CONCEPT OF MOTION

The scholastic approach to the study of motion was largely repudiated by both Galileo and Descartes, who sought to found a new science of motion on entirely different principles. Although there is no serious question that the scientific work of the seventeenth century owed a substantial debt to scholastic work on the subject, the self-concious rejection of Aristotelian principles is a recurrent theme in the scientific work of both Galileo and Descartes. The Galilean doctrines which are of greatest importance for our purposes are his claim that matter is indifferent to motion or rest, and the analysis of free fall in terms of uniform acceleration.[6] Descartes's most important contributions on this head include his refinement of the Galilean principle of inertia and his statement of a mechanistic paradigm for natural philosophy.[7] These set the stage for the development of mechanics in the seventeenth century and are an essential part of the background to Berkeley's *De Motu*.

---

[6] For more on Galileo's work and the background to it, see Drake and Drabkin (1969), Koyré (1939), and McMullin (1967).

[7] See Costabel (1967), Gabbey (1980) and Gueroult (1980) for helpful accounts of the Cartesian program for physics.

Galileo's commitment to the Copernican system of astronomy forced him to attempt to construct a physics for a Earth supposed to be in motion, since the Copernican model could hardly be upheld without supplying a physical science to explain the behavior of objects near the surface of the moving Earth. In seeking a new physics Galileo rejected the Aristotelian doctrine that every motion requires the continued action of a mover and settled instead on a quasi-inertial conception of motion in which a body is indifferent between motion and rest. As it turns out, Galileo's conception of inertia was considerably different from the Newtonian law of inertia. Indeed, Galileo propounded what we might call the "principle of circular inertia," which holds that a body at rest will remain at rest while a body set in circular motion will remain in circular motion unless acted upon by an outside force.[8] Acceptance of the principle allows Galileo to argue for a thesis of relativity in which an object is in motion only with respect to a system of bodies which are assumed to be at rest. Because all objects on the surface of the Earth participate in the Earth's diurnal rotation, a ball dropped from a tower will rotate with the tower and strike at its base. To an observer in the tower the ball will seem to fall straight down; and yet an imaginary observer in space would describe the ball's motion as a curve composed of downward movement of the ball and its rotation along with the Earth and tower.

Galileo's famous analysis of free fall in his *Two New Sciences* led to the formulation of the law of falling bodies: free fall is uniformly accelerated motion in which the distance traversed is as the square of the time. When applied to the case of projectile motion Galileo's account of free fall treats the horizontal and vertical components of the projectile's motion independently, with the result that a projectile will travel in a parabolic path determined by the constant gravitational acceleration and the initial force imparted to it. Moreover, Galileo argued that, ignoring air resistence, the acceleration of gravity is both constant and independent of the mass of the accelerated object.

Another respect in which Galileo's work departed from the Aristotelian and scholastic tradition was his refusal to attempt a causal explanation of such "natural motions" as gravitational acceleration, while

---

[8] See Galilei (1953, 147) for a concise statement of this principle.

stressing the importance of a mathematical analysis of motion. Galileo's insistence upon the primacy of mathematics in the understanding of nature is most clearly set forth in his manifesto *The Assayer*, when he makes his famous declaration that

> Philosophy is written in this grand book, the universe, which stands continually open to our gaze. But the book cannot be understood unless one first learns to comprehend the language and read the letters in which it is composed. It is written in the language of mathematics, and its characters are triangles, circles, and other geometric figures without which it is humanly impossible to understand a single word of it; without these, one wanders about in a dark labyrinth. (Galilei 1957, 237-8)

Parallel to this insistence upon a mathematization of the physics of motion is Galileo's advice to abandon the search for physical causes. Unlike Aristotelians who had attempted to explain the descent of heavy bodies in terms of the action of a quality called "gravity" or the influence of a substantial form, Galileo insists that we must content ourselves with mathematically exact descriptive laws of motion and leave causal inquiry behind. Thus, in the Second Day of the *Dialogue concerning the Two Chief World Systems* the Aristotelian Simplicio insists that the cause of the descent of heavy bodies is well known: "everybody is aware that it is gravity." To this, Galileo's spokesman Salviati replies:

> You are wrong, Simplicio; what you ought to say is that everyone knows that it is called "gravity." What I am asking you for is not the name of the thing, but its essence, of which essence you know not a bit more that you know about the essence of whatever moves the stars around. . . . [W]e do not really understand what principle or what force it is that moves stones downward, any more that we understand what moves them upward after they leave the thrower's hand, or what moves the moon around. (Galilei 1953, 234)

Similarly, in the *Two New Sciences* Salviati insists that the law of falling bodies should be developed mathematically and compared against experiment, but that the cause of acceleration can be left unanalyzed:

> The present does not seem to me to be an opportune time to enter into the investigation of the cause of the acceleration of natural motion, concerning which various philosophers have produced various opinions, some of them reducing this to approach to the center; others to the presence of successively less parts of the medium [remaining] to be divided; and others to a certain extrusion by the surrounding medium which, in rejoining itself behind the moveable, goes pressing and continually pushing it out. . . . For the present, it suffices. . . to investigate and demonstrate some attributes [*passiones*] of a motion so accelerated (whatever be the cause of its acceleration) that the momenta of its speed go increasing, after its departure from the rest, in that simple ratio with which the continuation of time increases, which is the same as to say that in equal times, equal additions of speed are made. And if it shall be found that the events that then shall have been demonstrated are verified in the motion of naturally falling and accelerated heavy bodies, we may deem that the definition assumed includes that motion of heavy things, and that it is true that their acceleration goes increasing as the time and the duration of motion increases. (Galilei 1974, 158-9)

As we will see, Berkeley's conception of physics bears a striking resemblance to these Galilean doctrines. He, too, upholds the importance of a mathematical treatment of physical phenomena, but warns that the search for true causes can only lead away from physics and into metaphysics. In fact, Berkeley goes to the extreme of claiming that physics cannot discover true causes, but that the search for genuinely active principles must be left to metaphysics.

The natural philosophy of Descartes is dominated by a conviction that the true principles of metaphysics are necessary for the development of a genuine science of motion, and in this respect he differs sharply from Galileo's professed agnosticism on the metaphysical issues underlying his science. Descartes frequently insists that a properly developed physics must have its roots in metaphysical principles which are known with certainty and suffice to derive all of the phenomena of nature. In the Cartesian scheme, the concepts of matter and motion are fundamental to the explanation of the phenomena of nature, and these can in turn be grasped by philosophical meditation. In his *Principles of Philosophy* Descartes attempted to work out his program for physics, basing the entire scheme upon metaphysical principles. He notoriously declared that the essence of body is extension, and his analysis of motion commits him to a strong relativity thesis in which a body is only in motion with respect to others which are regarded as at rest, so that in Article 25 of the second part of the *Principles* he defines motion as

> *the transfer of one piece of matter, or one body, from the vicinity of the other bodies which are in immediate contact with it, and which are regarded as being at rest, to the vicinity of other bodies.* (*AT*, 9B: 53; *CSM*, 1: 233)

Since the essence of body is extension, the motions of bodies cannot be explained in terms of substantial forms or occult qualities. And because the nature of extension is investigated by the science of geometry, Descartes can assert that

> The only principles which I accept, or require, in physics are those of geometry and pure mathematics; these principles explain all natural phenomena, and enable us to provide quite certain demonstrations regarding them. (*AT*, 9B: 78; *CSM*, 1: 247)

Further, Descartes holds that no explanation is required for why bodies in motion remain in motion; he assumes that both motion and rest are states to which a body is indifferent. In his first two laws of motion, Descartes explicitly states an inertial principle in which rest and uniform rectilinear motion are equally natural states of a body:

> The first law of nature: each and every thing, in so far
> as it can, always continues in the same state; and thus
> what is once in motion always continues to move.
> The second law of nature: all motion is in itself recti-
> linear; and hence any body moving in a circle always
> tends to move away from the center of the circle which
> it describes. (AT, 9B: 62-63; CSM, 1: 240-1)

Descartes set forth a complex system of laws of impact which were intended to account for all natural motions. Unfortunately for Descartes,
these laws are seriously flawed and cannot serve as the basis of an adequate physics, but his general procedure marks an important development in the mechanistic program for physics.

In addition to his principle of inertia, Descartes also propounded a
conservation law which forms an important part of the background to
Berkeley's *De Motu*. Descartes declared God to be the first cause or
principle of all motions in nature, and from the immutability of God he
derived the principle that the "quantity of motion" in the universe must
remain constant:

> God is the primary cause of motion; and he always
> preserves the same quantity of motion in the universe.
> (AT, 9B: 61; CSM, 1: 240)

Quantity of motion here is measured by the product of mass and speed,
or undirected velocity. The principle of conservation of momentum is
familiar today, except that velocity is conceived as a vector quantity,
so that in a system of bodies with masses m_i and velocities $\vec{v}_i$ the sum
of the products, $\sum m_i \vec{v}_i$, remains constant. Much of the problem with
Descartes's laws of impact can be traced to his treatment of velocity as a
scalar quantity, and the history of physics from Descartes to Newton was
dominated by attempts to find the appropriate form of the basic laws
of motion. For Descartes, the sum $\sum m_i |\vec{v}_i|$ of masses and *undirected*
velocity is conserved, while the direction but not the quantity of motion
can be altered.

1.3 LEIBNIZ AND THE PHYSICS OF FORCES

The Cartesian dictum that the essence of body consists solely in extension was challenged by Leibniz, who sought to ground physics in the consideration of forces rather than bare extension. Leibniz agreed with the Cartesian methodology which holds that physical principles must be grounded in metaphysical truths, but he took the inadequacies of Cartesian physics (such as its false laws of impact) as symptomatic of deeper problems in Descartes's metaphysics. Leibniz asserts over and over again that the simple concept of extension is insufficient to account for all the properties of bodies and must be supplemented by the consideration of forces. This introduction of forces into the ontology of physics results in a new science of dynamics. In setting out his own views on this subject Leibniz developed a complex labyrinth of concepts, mixing physical notions such as resistance with metaphysical concepts such as substance. He also introduces a bewildering array of technical terms in the course of his treatment of dynamics, which led Berkeley to charge that he had burdened physics with a confusing collection of empty abstract names that served to explain nothing. Berkeley was familiar with Leibniz's presentation of his dynamics in the paper *Specimen Dynamicum*, and we must therefore outline some of the key aspects of this work.

One important feature of Leibniz's doctrines in the *Specimen Dynamicum* (and indeed of his philosophy generally) is his interest in reconciling Aristotelian doctrines with the tenets of "the moderns." In particular, he holds that the Aristotelian conception of substantial form or entelechy is important for a true metaphysics, although he grants that it cannot be appealed to in the explanation of particular phenomena. The Cartesian rejection of substantial forms in favor of a conception of body as pure extension is, on Leibniz's view, the source of serious error in both physics and metaphysics, as he explains early on in the *Specimen Dynamicum*:

> Elsewhere we urged that in corporeal things there is something over and above extension, in fact, something prior to extension, namely, that force of nature implanted everywhere by the Creator. This force does not consist in a simple faculty, with which the schools

seem to have been content, but is further endowed with *conatus* or *nisus*, attaining its full effect unless it is impeded by a contrary *conatus*. . . . But if we should not attribute this *nisus* to God, acting by miracle, then it is certainly necessary that he produce that force in bodies themselves, indeed, that it constitute the innermost nature of bodies, since to act is the nature of substances, and extension means nothing but the continuity or diffusion of an already presupposed striving and reacting (that is, resisting) substance. . . . Just as our age has already saved from scorn Democritus' corpuscles, Plato's ideas, and the Stoics' tranquility in the light of the most perfect interconnection of things, so now we shall make intelligible the teachings of the Peripatetics concerning forms or entelechies, notions which seemed enigmatic for good reason, and were scarcely perceived by their own authors in the proper way. (*GM*, 6: 235; *AG*, 118)

The technical terms '*nisus*' and '*conatus*' in this passage derive from the Latin verbs *nitor* and *conor*, which indicate straining, effort, or endeavor. The science of dynamics will then introduce active forces into bodies, literally animating the Cartesian universe of bare extension.

Leibniz's account of force in the *Specimen Dynamicum* begins with a four-fold distinction among different kinds of forces: force can be either active or passive, and again either primitive or derivative. We thus get a classificational schema of primitive active force, derivative active force, primitive passive force, and derivative passive force. Leibniz then interprets these distinctions within the context of an Aristotelian theory of motion:

Active force (which might not inappropriately be called *power* [*virtus*] as some do) is twofold, that is, either *primitive*, which is inherent in every corporeal substance *per se*. . . , or *derivative*, which, resulting from a limitation of primitive force through the collision of bodies with one another, for example, is found in different degrees. Indeed, primitive force (which is nothing but the first entelechy) corresponds to the *soul* or *substantial form.* . . . Similarly, passive force is also twofold, either primitive or derivative. And indeed, the *primitive force of being acted upon* [*vis primitiva patiendi*] or of *resisting* constitutes that which is called *primary matter* in the schools, if correctly interpreted. This force is that by virtue of which it happens that a body cannot be penetrated by another body, but presents an obstacle to it, and, at the same time is endowed with a certain laziness, so to speak, that is, an opposition to motion, nor further, does it allow itself to be put into motion without somewhat diminishing the force of the body acting on it. As a result, the *derivative force of being acted upon* later shows itself to different degrees in *secondary matter*.[9] (*GM*, 6: 236-7; *AG*, 119-120)

The key task for the science of dynamics is to study the derivative forces present in collision and the resistence offered by various bodies. In the course of this project Leibniz draws further distinctions, first between *conatus* and *impetus*, then between *motio* and *motus*, thirdly between elementary *nisus* and actual *nisus*, and lastly between dead and living forces. Because these concepts are specifically mentioned in Berkeley's *De Motu*, it is important that they be outlined here.

Leibniz defines *conatus* as "Velocity taken together with direction," and contrasts it with *impetus*, which is "the product of the bulk [*moles*] of a body and its velocity." (*GM*, 6: 237; *AG*, 120) Thus, the *conatus* of a body will be its directed velocity, while the *impetus* will be the same as the Cartesian "quantity of motion," namely the product $m|\vec{v}|$ of mass and the scalar magnitude of the directed velocity. Leibniz understands

---

[9] The references here to "primary matter" and "secondary matter" are to scholastic metaphysical doctrines. Primary matter is a "pure potency" devoid of form which underlies all bodies. Any existing body will, however, be composed of both form and matter, and such composites are "secondary matter" or the bodies of ordinary experience.

the velocity of a body to be the velocity at an instant in time, which leads him to distinguish the body's motion at an instant (*motio*) from its motion over time (*motus*): "just as we can distinguish the present descent from descent already made, descent which it augments, so too we can distinguish the present or instantaneous element of motion [*motus*] from that same motion extended through a period of time, and call the former *motio*." (*GM*, 6: 237; *AG*, 120)

This talk of "elementary parts" or instantaneous increments extends to the theory of *nisus*. The elementary *nisus* or "solicitation" of a body just set in motion is an infinitesimal part of its *impetus*, while the actual *nisus* (which is the same as *impetus*) is an infinite sum of these infintesimal instantaneous increments. As Leibniz puts it: "the *nisus* is twofold, that is, elementary or infinitely small, which I also call *solicitation*, and that which is formed from the continuation or repetition of elementary nisus, that is, the *impetus* itself." (*GM*, 6: 238; *AG*, 121) This theory can be illustrated by an example. Imagine a chandalier hanging from a wire which suddenly breaks. When the wire breaks, the chandalier goes from a state of rest to a state of motion and potential energy is converted into kinetic energy. But at the very instant of the breaking, Leibniz conceives the motion to begin with the "solicitation of gravity," or the product of mass and an infinitesimal motion. As it descends to the floor, the chandalier continuously accumulates these infinitesimal increments and accelerates to land with a crash.

A similar distinction applies to forces, which can be either living or dead. The dead force is an instantaneous increment of a living force, as Leibniz explains:

> From this it follows that *force* is also twofold. One force is elementary, which I also call *dead force*, since motion [*motus*] does not yet exist in it, but only a solicitation to motion [*motus*], as with . . . a stone in a sling while it is still being held in by a rope. The other force is ordinary force, joined with actual motion, which I call *living force*. An example of dead force is centrifugal force itself, and also the force of heaviness [*vis gravitatis*] or centripetal force, and the force by which a stretched elastic body begins to restore itself. But when we are

dealing with impact, which arises from a heavy body which has already been falling for some time, or from a bow that has already been restoring its shape for some time, or from a similar cause, the force in question is living force, which arises from an infinity of continual impressions of dead force. (*GM*, 6: 238-9; *AG*, 121-2)

There is a strong analogy here between Leibniz's conception of force and motion as consisting of an infinite collection of infinitesimal elements and his treatment of the calculus, where geometric magnitudes are composed of infinite collections of infinitesimals. This is no accident, since one of the key tasks of the calculus is to provide the mathematical background for the study of motion.

Leibnizian dynamics is one of the principal targets of Berkeley's *De Motu*, and especially the distinction between living and dead forces, or between impetus and solicitation. As we will see, Berkeley finds the entire Leibnizian enterprise to be founded upon a series of false abstractions which introduce meaningless terms into the language of physics.

1.4 NEWTONIAN MECHANICS

Newton's *Principia* is the other principal target of Berkeley's *De Motu*, particularly for its doctrine of universal gravitation and the distinction between absolute and relative space. Although the elements of the Newtonian system are familiar, they deserve a summary here because of their importance as part of the context of Berkeley's work.[9] Newton's presentation of his physics is modeled on the Euclidean treatment of geometry and begins with definitions and fundamental axioms or laws of motion from which natural phenomena are to be derived. For our purposes, the most important of these definitions concern motive forces and the doctrine of absolute space.

---

[9] Among the many studies of Newton, Cohen (1980), Koyré (1965), Scheurer and Debrock (1988), Westfall (1971), and Westfall (1980) deserve mention.

Newton's introduction of the concept of force reflects the important link between inertia and acceleration. For Newton, every body contains an "innate force" by which it seeks to remain in a state of rest or uniform rectilinear motion. Anything which intereferes with the inertial state of a body (such as gravitation or impact) produces an "impressed force." These concepts are defined in the third and fourth definitions of the *Principia*. Definition III reads:

> *The* vis insita *or innate force of matter, is a power of resisting, by which every body, as much as in it lies, continues in its present state, whether it be of rest, or of moving uniformly forwards in a right line.* (*Principia*, 1: 2)

In explicating this definition Newton observes that "this *vis insita* may, by a most significant name, be called inertia (*vis inertiæ*) or force of inactivity." This force of inertia is contrasted with "impressed forces" in the fourth definition, which declares an impressed force to be "*an action exerted upon a body, in order to change its state, either of rest, or of uniform motion in a right line.*" (*Principia*, 1: 2)

Among impressed forces the most important are those which direct a body toward a central point. These "centripetal" (literally, "center-seeking") forces include the force of gravitation, and are defined and explicated in Definition V:

> *A centripetal force is that by which bodies are drawn or impelled, or any way tend, towards a point as to a centre.*
> Of this sort is gravity, by which bodies tend to the centre of the earth; magnetism, by which iron tends to the loadstone; and that force, whatever it is, by which the planets are continually drawn aside from the rectilinear motions, which otherwise they would pursue, and made to revolve in curvilinear orbits. . . . [Bodies in orbits] all endeavor to recede from the centres of their orbits; and were it not for the opposition of a contrary force

which restrains them to, and detains them in their orbits, which I therefore call centripetal, would fly off in
right lines, with an uniform motion. (*Principia*, 1: 2-3.)

These definitions clearly endorse the inertial conception of motion set
forth by Galileo and Descartes, and with them Newton lays much of the
foundations of classical physics.

Where Galileo and Descartes had stressed the relativity of motion,
Newton made a radical distinction between relative or common notions
of space and time and the absolute or true quantities. In the famous
"Scholium" to the Definitions in the *Principia* Newton defines absolute
time, absolute space, absolute place and absolute motion. In his words:

I. Absolute, true, and mathematical time, of itself, and
from its own nature, flows equably without relation to
anything external, and by another name is called duration: relative, apparent, and common time, is some
sensible and external (whether accurate or unequable)
measure of duration by the means of motion, which is
commonly used instead of true time; such as an hour,
a day, a month, a year.
II. Absolute space, in its own nature, without relation
to anything external, remains always similar and immovable. Relative space is some movable dimension or
measure of the absolute spaces; which our senses determine by its position to bodies; and which is commonly
taken for immovable space; such is the dimension of
a subterraneous, and aerial, or celestial space, determined by its position in respect of the earth. . . .
III. Place is a part of space which a body takes up, and
is according to the space, either absolute or relative. I
say, a part of space; not the situation, nor the external
surface of the body. . . .
IV. Absolute motion is the translation of a body from
one absolute place into another; and relative motion,
the translation from one relative place into another.
Thus in a ship under sail, the relative place of a body

is that part of the ship which the body possesses; or
that part of the cavity which the body fills, and which
therefore moves together with the ship; and relative
rest is the continuance of the body in the same part of
the ship, or of its cavity. But real, absolute rest, is the
continuance of the body in the same part of that immovable space, in which the ship itself, its cavity, and
all that it contains, is moved. (*Principia*, 1: 6-7)

Newton's insistence upon this distinction was motivated in part by the
desire to give his laws of motion a kind of objectivity which is lacking in
relativistic accounts of motion such as that of Descartes. In the Cartesian system no body has a determinate velocity, but Newton desired to
develop a physics where it would be an objective fact that, for example,
the Earth is in motion.

Newton was not content simply to state his distinction between absolute and relative motions, but thought that the existence of an absolute
reference frame could be demonstrated by physical experiments. In particular, his famous "bucket argument" attempts to show that absolute
motion can be detected by measuring forces arising from circular motions, since "[t]he effects which distinguish absolute from relative motion
are, the forces of receding from the axis of circular motion." (*Principia*, 1: 10) Newton argues as follows: take a bucket filled with water
and suspend it from a cord that has been twisted. Release the bucket
and it will begin to rotate; before the bucket was released, the surface of
the water was flat and the bucket and water were not in motion relative
to one another. Shortly after being released, the bucket will be spinning
rapidly relative to the water, but the surface of the water will remain
flat. Eventually the water will acquire circular motion and ascend the
side of the bucket, making the surface of the water concave. Finally, the
water will be at rest with respect to the spinning bucket, but the surface
will be concave. The concavity shows the presence of genuine motion,
for even though the water and bucket are not in motion relative to one
another, the distortion of the surface shows that forces are applied to
the water and it must therefore be in motion. As Newton puts it:

This ascent of the water shows its endeavor to recede from the axis of its motion; and the true and absolute circular motion of the water, which is here directly contrary to the relative, becomes known, and may be measured by this endeavor. At first, when the relative motion of the water in the vessel was greatest, it produced no endeavor to recede from the axis; the water showed no tendency to the circumference, nor any ascent towards the sides of the vessel, but remained of a plain surface, and therefore its true circular motion had not yet begun. But afterwards, when the relative motion of the water had decreased, the ascent thereof towards the sides of the vessel proved its endeavor to recede from the axis; and this endeavor showed the real circular motion of the water continually increasing, till it had acquired its greatest quantity, when the water rested relatively in the vessel. And therefore this endeavor does not depend upon any translation of the water in respect of the ambient bodies, nor can true circular motion be defined by such translation. (*Principia*, 1: 10-11)

Newton presents a similar argument in the form of a thought experiment involving the behavior of two globes connected by a cord. We can imagine them rotating about their common center of gravity, thereby inducing tension in the cord. But if we imagine these two globes to be the only bodies in the universe, we can still distinguish their state of rest from the state of circular motion by measuring the tension in the cord. Thus, according to Newton, the distinction between absolute rest and absolute motion can be upheld, even if absolute space and time are insensible in themselves.

Given these fundamental definitions and distinctions, Newton set out his famous "Axioms, or Laws of Motion." These read:

Law I. *Every body continues in its state of rest, or of uniform motion in a right line, unless it is compelled to change that state by forces impressed upon it.*

Law II. *The change of motion is proportional to the motive force impressed; and is made in the direction of the right line in which that force is impressed.*

Law III. *To every action there is always opposed an equal reaction; or, the mutual actions of two bodies upon each other are always equal, and directed to contrary parts.* (*Principia*, 1: 13)

These are then supplemented by the principle of universal gravitation, which Newton puts forward in Book III of the *Principia* after a laborious argument to establish "by induction" that the observed behavior of celestial and terrestial bodies shows the presence of a common gravitating force. In Proposition VII, Theorem VII of Book III, Newton announces

That there is a power of gravity pertaining to all bodies, proportional to the several quantities of matter which they contain. (*Principia*, 2: 414)

Taken together Newton's definitions, laws of motion, and the principle of universal gravitation constitute a universal system of physics in which the motions of both planets and billiard balls can be predicted and explained on the basis of a small number of common principles. The empirical success of Newton's theory posed something of a problem for Berkeley, whose epistemological and metaphysical principles mandate the rejection of nearly all of Newton's basic assumptions. Berkeley obviously cannot accept the concpet of an inertial force in bodies, proportional to their quantities of matter, neither can he allow a physics based upon absolute space and time, nor do his principles allow for a universal gravitational force.

2. DISPUTED POINTS IN THE MECHANICAL PHILOSOPHY

It would be a serious mistake to imagine that the history of natural philosophy in the seventeenth and eighteenth centuries is that of a steady and uninterrupted march from Aristotelian notions to the universally accepted Newtonian system. At every stage in the development of physical theories there were unresolved issues and points of significant dispute. Indeed, disputed and controversial issues outnumber points of general agreement in the history of physics from Galileo to Newton. Several disputes are of particular importance for a reading of *De Motu*, because Berkeley makes frequent reference to them and attempts to show that lack of consensus in physics is the product of unsound metaphysical and methodological principles. The controversies of interest for a reading of Berkeley are three: the "*vis viva*" controversy between Leibniz and the Cartesians, the doctrine of the infinite force of percussion, and disputes over the nature of gravitation.

2.1 THE *VIS VIVA* CONTROVERSY

In 1686 Leibniz published an article in the journal *Acta Eruditorum* entitled "A Brief Demonstration of a Notable Error of Descartes and Others concerning a Natural Law. . ."[10] In it he argued that the Cartesian principle of the conservation of motion (mv) was mistaken, and proposed to replace it with a law of conservation of *vis viva* or living force, represented by the quantity mv^2. The inadequacy of Descartes's laws of impact was well known before Leibniz attacked the conservation principle, but there had been no disagreement with the fundamental idea that the universe contained a constant "quantity of motion." Indeed, by treating velocity as a vector quantity and modifying their laws of impact, the Cartesians upheld the law of conservation of momentum which is a familiar part of classical mechanics. Leibniz's argument in the "Brief Demonstration" is simple enough, but Cartesians refused to accept some of its basic suppositions and the result was a prolonged controversy over the proper formulation of a basic conservation law.

The argument of the "Brief Demonstration" begins with two principles which Leibniz takes to be accepted by all Cartesians, "as well as other philosophers and mathematicians of our times." These include:

(I) A body falling from a certain height acquires just that force necessary to raise it to that height again.

---

[10] The piece was published in the *Acta Eruditorum* for March of 1686. It can be found in *GM*, 6: 117-19.

(II) The same force is required to raise a body of four
lb. one yard as to raise a body of one lb. four yards.
(GM, 6: 117)

Now assume a body A with a unit mass to fall from a height of four
yards, and a body B with four times the mass to fall from one yard.
By principle (I), both bodies will acquire just enough force to raise
themselves back to their original heights of four and one yards. But
by principle (II) the force acquired by A will be sufficient to raise B to
its original height, and *vice versa*. However, Galileo's analysis of free
fall shows that the distance travelled is proportional to the square of
the time, so that the velocity is proportional to the square root of the
distance. Then the velocity of A will be proportional to two, while the
velocity of B is proportional to one. Multiplying mass times velocity in
each case shows that the "quantity of motion" of B must be twice that
of A; and yet the quantity of force generated in each case must be the
same. Leibniz concludes that mv is simply not the appropriate measure
of the quantity of force, and argues that mv^2 or *vis viva* must be the
conserved quantity.

Responses to this Leibnizian argument took many forms and need
not be dealt with in any detail here.[11] In general, Cartesians defended
the law of conservation of mv by arguing that the true measure of the
effect produced by the descent of the bodies A and B was not simply the
height to which they could raise another body, but that the *time* taken
to produce the effect must be taken into account. If we regard the time
required to produce an effect (in essence, considering the velocity with
which a descending body can raise another body), the conservation of
mv can be vindicated.

In an important sense the root of the controversy is the question
of what should be taken as the proper measure of the effect produced
by the motion of bodies. This dispute dragged on for several decades
and was still a matter of concern when the Paris Academy proposed
its prize essay competition in 1720. For Berkeley, the matter can be
resolved easily by avoiding the abstraction of velocity from force. As

---

[11] See Hankins (1965), Iltis (1970), Iltis (1971), Laudan (1968), and Papineau
(1977) for differing opinions of the controversy and its eventual resolution.

he argues in §15, Leibniz's opinion "supposes the force of a body to be distinguished from momentum, motion, and impetus, and it collapses when this supposition is removed." This is unfortunately not a terribly enlightening remark, since it would seem to apply as well to those who measure force through the product mv.

2.2 THE FORCE OF PERCUSSION

Another issue which attracts Berkeley's attention in *De Motu* is the seemingly paradoxical opinion that the force of percussion is infinite. A principal source of this doctrine is Galileo's *Two New Sciences*, which contains a dialogue on the relationship between the force of gravitation and the force of percussion.[12] The problem arises when we compare the force produced by a dead weight with that produced by the impact of a moving body. To take Galileo's example, we can compare the effect of a pile driver with that of a dead weight placed upon a pole. Imagine that the pile driver drops a weight of 100 pounds and drives a pole four inches into the ground, while a dead weight of 1000 pounds would produce the same effect. Then consider a second impact from the pile driver, again delivering 100 pounds from the same height, and suppose that the pole is driven two inches further into the ground.

The problem is now to measure the force of the impact against the force of the dead weight, taking into account both the first and second impacts of the pile driver. Galileo's spokesman Salviati asks the interlocutor Aproino whether the second impact of the pile driver (which drove the pole a further two inches) would produce an effect comparable to that of the 1000 pounds of dead weight, which leads to the following exchange:

> *Salv.* Must we suppose that [the pole] would have been driven as much by the pressure of that same dead weight?
> *Apr.* So it seems to me.

---

[12] The "Sixth Day" dialogue which treats the force of percussion was not published until 1718, but Galileo and his associates had clearly discussed the problem in detail. See Moscovici (1968) and the "Preface" to Borelli (1667).

Salv. Alas, Paolo, for us; this must be emphatically denied. For if in the first placement, the dead weight of 1000 pounds drove the pole only four inches and no moe, why will you have it that by merely being removed and replaced, it will drive the pole two more inches? Why did it not do this before it was removed, while it was still pressing? Do you suppose that just taking it off and gently replacing it makes it do that which it could not do before?

Apr. I can only blush and admit that I was in danger of drowning in a glass of water. (Galilei 1974, 287)

This line of thought is then pursued by Sagredo, who argues that the force of percussion must then be infinite:

Sagr. Already I seem to understand that the truth may be that the force of impact is imense, or infinite. For in the above experiment, given that the first blow will drive the pole four inches and the second, three, and continuing ever to encounter firmer ground, the third blow will drive it two inches, the fourth an inch and one-half, the ensuing ones a single inch, one-half, one-fourth, and so on; it seems that unless the resistance of the pole is to become infinite through this firming of the ground, the repeated blows will always budge the pole, but always through shorder and shorter distances. But since the distance may become as small as you please, and is always divisible and subdivisible, entrance [of the pole] will continue; and this effect having to be made by the dead weight, each [movement] will require more weight than the preceeding. Hence it may be that in order to equal the force of the latest blows, a weight immensely greater and greater will be required. (Galilei 1974, 288)

Eventually Galileo concludes that the smallest impact produces an effect infinitely greater than that of a dead weight, which acts by simple gravitation without motion.

This result seemed deeply paradoxical to many, although it did not lead to a prolonged dispute such as that over the Leibnizian concept of *vis viva*. Torricelli, for example, devoted three of the seven lectures in his *Lezione Accademiche* to the force of percussion. In his introductory lecture he declares:

> The force then of percussion (that on which we will first discourse) bears, in my opinion, the crown of the principate among the scene of wonders. This is because it is the most striking of all the discoveries of mechanics, and is perhaps the most recondite, and the most abstruse among all the arcana of nature.[13]

Torricelli was not alone in his interest in the problem. Giovanni Alfonso Borelli considered it in his book *De Vi Percussionis* (Borelli, 1667), and it was a standard problem in many expositions of seventeenth-century physics.

Leibniz approached the problem of the force of percussion by way of his theory of elementary nisus or solicitation of gravity. On Leibniz's account, the force of percussion is a living force, which is infinite in comparison to an elementary nisus such as the solicitation of gravity. Then the paradox of the infinite force of percussion can be resolved: the force of percussion, although finite, is infinite in comparison to the solicitation of gravity. A body acted upon by gravity, but hindered from moving, can produce an effect which is infinitesimal with respect to a moving body. Berkeley's discussion in *De Motu* shows that he had read Torricelli, Borelli, and Leibniz on this issue. Moreover, he found the whole discussion to be founded on the mistake of abstracting forces from their effects. As he puts it in §10:

---

[13] "La forza poi della Percossa (sopra la quale faremo questo discorso) porta a mio giudizo nella scena della maraviglie la corona del Principato. Questa per esser la più efficace fra tutte le invenzioni della Meccanica, è forse più recondito, e il più astruso fra tutti gli arcana della Natura." (Torricelli 1715, 3)

And yet it must be allowed that no force is immediately
felt by itself, nor otherwise known and measured except
by its effect; but there is no effect of a dead force or
of a simple gravitation in a quiescent body subject to
no actual change. There is, however, some effect of
percussion. Since, therefore, forces are proportional to
effects, we may conclude that there is no dead force.
But neither should we conclude on that account that
the force of percussion is infinite. For it is not proper
to take any positive quantity as infinite on the grounds
that it exceeds by an infinite ratio a null quantity or
nothing. (*De Motu*, §10)

This characteristically Berkeleyan rejection of the infinite goes to the
heart of the matter and reflects the standpoint which eventually led to
the downfall of the doctrine of the infinite force of percussion.[13]

2.3 THE NATURE OF GRAVITATION

Newton's *Principia* was hailed in England as the complete system of
natural philosophy, but the reception on the Continent was considerably
less favorable.[14] In particular, Newton's principle of universal gravitation was rejected by many who accused Newton of relying on an "occult
quality" in bodies to explain the phenomena of motion. Cartesians and
those who worked in the Cartesian tradition (notably Huygens and Leibniz) favored a "vortex model" of mechanics in which attractive forces
arise from the steady flow of fine particles. A vortex theory of planetary
orbits, for example, treats the planets as swept along by a current of
aetherial particles rather like leaves in the wind.[15]

---

[13] See Moscovici (1968) for an account of the problems involved in the infinitesimal analysis of percussion and the decline of the Galilean analysis of the
phenomenon.
[14] See Guerlac (1981) for a study of the reception of Newtonianism on the
Continent.
[15] See Aiton (1972) on vortex theory and its history.

Newton embarrassed the vortex theory by showing that celestial phenomena are inconsistent with the vortex hypothesis. In particular, Kepler's laws of planetary orbit cannot be satisfied by vortical motion, while the highly eccentric orbits of comets would seem to cut through the vorticies without disturbing the orbit of planets. Newton concluded that interplanetary space is a pure vaccuum, with gravitational attraction acting across the void. The difficulty with this view is that it violates the fundamental principle of mechanism, that of "action by contact." The Newtonian system depends upon action at a distance, which many held to be simply unintelligible. After all, it is difficult to explain how one body can act upon another without coming into contact with it or acting through an intervening medium. And yet the empirical success of the theory of universal gravitation makes it a highly attractive alternative when compared to the discredited vortex theories. The chief result of this dilemma was an extended series of disputes between Newtonians and their Continental counterparts over the nature of gravitation, but also over the basic principles of scientific methodology. The most notable of these was the famous exchange of letters between Leibniz and Samuel Clarke (a disciple of Newton), but there were several others.[16]

Roger Cotes, Plumian Professor of Astronomy at Cambridge, addressed these issues in his preface to the second edition of the *Principia*. After setting out the evidence that both terrestial and celestial bodies act in accordance with the principle of universal gravitation, and that no exception is known, Cotes concludes that we must acknowledge gravitation as an inherent property of matter. He then adds:

> Some I know disapprove this conclusion, and mutter
> something about occult qualities. They continually are
> cavilling with us, that gravity is an occult property and
> occult causes are to be quite banished from philosophy.
> But to this the answer is easy: that those are indeed
> occult causes whose existence is occult, and imagined
> but not proved; but not those whose real existence is

---

[16] See Alexander (1956) on the Leibniz-Clarke dispute. Other studies of these conflicts include Burtt (1950), Iltis (1973a), Iltis (1973b), and several of the essays in Koyré (1965).

clearly demonstrated by observations. Therefore gravity can by no means be called an occult cause of the
celestial motions, because it is plain from the phenomena that such a power does really exist. Those rather
have recourse to occult causes, who set imaginary vor-
tices of a matter entirely fictitious and imperceptible
by our senses, to direct those motions. (*Principia*, 1:
xxvi-xxvii)

Of course this merely sidesteps the crucial issue, which is to explain how
the force gravitation can act at a distance. Many Newtonians professed
agnosticism on this question and were content to echo Newton's famous
dictum, "I have not been able to discover the cause of those properties
of gravity from phenomena, and I frame no hypotheses." (*Principia* 2:
547)

Berkeley was impressed with the success of the Newtonian theory,
but equally disturbed by its apparent reliance upon an occult quality
in the explanation of natural phenomena. His resolution of the problem
was to urge that physics should not concern itself with causal inquiry,
but merely with the articulation of general principles which can be used
to predict phenomena:

[I]t is the concern of the physicist or mechanician to
consider only the rules, not the efficient causes, of impulse or attraction, and, in a word, to set out the laws
of motion: and from the established laws to assign the
solution of a particular phenomenon, but not an efficent
cause. (*De Motu* §35)

In Berkeley's view, real active causes are the proper subject of metaphysics. Physics should confine itself to the discovery of laws which
have high predictive value, but the search for true causes is in vain.
This doctrine echoes Berkeley's metaphysical thesis that only minds are
truly active, and what we ordinarily take to be a causal connection between events is merely the relation of a sign to the thing signified.

3. THE PLACE OF *DE MOTU* IN BERKELEY'S PHILOSOPHY

The principal task of *De Motu* is to set forth an interpretation of physical theory which is consistent with Berkeleyan epistemological and metaphysical principles. In particular, Berkeley continually stresses that the study of motion has been hindered by excessive abstraction and the attribution of causal powers to bodies. The critique of abstraction is, of course, a centerpiece of Berkeley's epistemology, and the doctrine that only minds are truly active is a familiar theme from his metaphysics. Perhaps surprisingly, there is no mention of immaterialism in *De Motu*. Berkeley certainly does not abandon immaterialism in this work, but his language is sufficiently vague to avoid the issue. He insists upon a distinction between mind (*mens*) and body (*corpus*) as two distinct classes of things, without the further claim that bodies do not comprise a class of extra-mental substances.

The perils of abstraction are critiqued throughout *De Motu*, with emphasis on two important cases. First, Berkeley argues that the doctrine of forces is the product of misguided abstraction; on this account, the postulation of mysterious forces behind observed motions results from trying to separate the idea of a force from the idea of motion or any other sensible quality. As a result, endless disputes (such as those over *vis viva* and the force of percussion) arise, and the opinions of natural philosophers run into absurdity. The second example of illegitimate abstraction is Newton's doctrine of absolute space. In this case, the attempt to abstract an idea of space from any sensible quality has led to the postulation of an infinite, immutable, invisible, *sui generis* object. But, as Berkeley puts it in §53, "all of its attributes are pirvative and negative: it seems therefore to be merely nothing."

Berkeley's insistence upon the causal inefficacy of bodies is a key element in his discussion of the "principle" of motion, and in particular in his dismissal of the idea that there is such a thing as force inherent in bodies. His argument for this point is very much in the tradition of Malebranche and French Occasionalists:[17]

---

[17] See Brykman (1979) for a study of the "Cartesianism" of *De Motu*.

All that which we know and have given the name 'body' contains nothing in itself which could be the principle or efficient cause of motion; for indeed impenetrability, extension, and figure include or connote no power of producing motion. On the contrary, reviewing singly not only these but any other qualities of body, whatever they might be, we will see that they are all in fact passive and there is nothing active in them which could in any way be understood as the source and principle of motion. (*De Motu*, §22)

Berkeley concludes that mind is the principle of motion, and that although the aim of physics is to discover the laws of motion, such laws cannot actually assign the true cause of motions.

Berkeley frequently linked *De Motu* to his other philosophical works, either explicitly or implicitly. Thus, in his critique of absolute space in §55, he refers back to §§110-117 of the *Principles of Human Knowledge*, and in the ninth "Query" at the end of *The Analyst* he advises the reader to "see a Latin treatise *De Motu*," to confirm that the doctrine of forces has involved mathematicians in disputes and paradoxes "over what they neither do nor can conceive." In his philosophical correspondence with the American Samuel Johnson, he sketches an account of natural philosophy which is quite close to that in the present work, and on the question of absolute space refers Johnson to "a Latin treatise, *De Motu*, which I shall take care to send you." (*Works*, 2: 280) In Dialogue VII of the *Alciphron*, Berkeley contends that physics has been burdened by the mysterious doctrine of forces, and although he does not actually cite *De Motu*, his choice of examples and his overall analysis remain unchanged. Even in *Siris*, which seems to promote a rather different conception of nature, Berkeley returns to the themes of *De Motu*:

And although a mechanical or mathematical philosopher may speak of absolute space, absolute motion, and of force as existing in bodies, causing such motion and proportional thereto; yet what these forces are which are supposed to be lodged in bodies, to be impressed on bodies, to be multiplied, divided, and communicated

from one body to another, and which seem to animate
bodies like abstract spirits or souls, hath been found
very difficult, not to say impossible, for thinking men
to conceive and explain; as may be seen by consulting
Borelli *De Vi Percussionis*, and Torricelli in his *Lezioni
Academiche*, among others. (*Works*, 5: 119)

Berkeley makes the link to *De Motu* explicit in the subsequent section
when he declares "it is very certain that nothing in truth can be measured or computed, beside the very effects of motion themselves," and
adds a footnote referring to "A Latin tract *de Motu*, published above
twenty years ago."

Critical opinions on *De Motu* have varied widely over the years.
A.A. Luce characterized it as "the application of immaterialism to contemporary problems of motion," and insisted that "apart from the *Principles* the *De Motu* would be nonsense.[18] Others have been less inter-
ested in immaterialism and have read the work as a statement of a kind
of proto-positivism in which Berkeley anticipates the doctrines of Ernst
Mach or the "verificationist" criterion of meaning for theoretical terms
in science.[19] Some have accused Berkeley of inconsistently rejecting the
doctrine of forces while endorsing Newtonian mechanics.[20] Still other
commentators have regarded him as holding that the theory of forces can
be replaced by talk about observed motions,[21] or see him as espousing
an instrumentalism in which scientific theories are acceptable for their
predictive value but not regarded as true.[22]

The place to start with an understanding of *De Motu* is to recognize
it as a contribution to eighteenth-century debates on the nature of force
and motion. Berkeley clearly thinks that the problems can be solved
by abandoning the abstractions in which Leibniz, Newton, and others

---

[18] These remarks appear in the "Editor's Introduction" to his edition of *De
Motu*, (*Works*, 4: 3-4).
[19] See Hinrich (1950), Popper (1953) and Myhill (1957) for this reading.
[20] Silver (1973) argues that Berkeley cannot accept Newtonian physics, while
Mirarchi ((1977a) tries to supply Berkeley with a concept of force which would
not mandate his rejection of mechanics.
[21] Brook (1973) takes a view along these lines.
[22] Buchdahl (1969) and Newton-Smith (1985) tend toward this reading.

have indulged. To this extent, *De Motu* can be read as an attempt to deliver on the promise of the *Principles* to free natural philosophy from burdensome false notions. Although it is consistent with the plan of the *Principles*, there is no reason to think that *De Motu* is purely and exercise in immaterialistic physics. As noted, Berkeley does not deny the existence of matter in this work and his conclusions could be accepted by anyone who agreed with the thesis of the causal inefficacy of bodies and the absurdity of abstraction; a Malebranchian occasionalist, for example, would find little or nothing to contest in it.

It is clear from the text that Berkeley accepted the physics of the early eighteenth century and regarded the mathematical analysis of natural phenomena as an appropriate procedure in natural philosophy. But it is equally clear that he denies the existence of forces, absolute space, and other key components of the Newtonian system. How can he do this consistently? I beleive that the answer lies in the final section of *De Motu*. There, Berkeley insists that

> Only by meditation and reasoning can truly active causes be brought to light from out of the enveloping darkness, and to some extent known. But to treat of them is the concern of first philosophy or metaphysics. And if to each science its province were allotted, its limits assigned, and the principles and objects which belong to it accurately distinguished, we could treat each with greater ease and perspicuity. (*De Motu*, §72)

The message is clear: there is a proper province for physics (articulation of the laws of nature), and another for metaphysics (which includes such "active causes" as the Divine and human minds). A physical theory may employ the language of cause and effect, but in such cases it does not speak truly. Following the "language model" of nature set forth in §§102-110 of the *Principles*, Berkeley sees the laws of nature as dealing with signs rather than causes; an astute natural philosopher is one who has mastered the language of nature and can make successful predictions about what will happen under various circumstances. Although the text of nature is no fiction, scientific theories which purport to describe causes are. Or, to recall Berkeley's insistence upon the heirarchy of sciences:

metaphysics and theology are the province of truth, while natural science tells us a useful story.

4. A NOTE ON THE TEXT AND TRANSLATION

The present edition is based on that contained in the *Miscellany*, published by Berkeley in 1752. The very few material variants between this text and the first edition of 1721 are noted, and some corruptions in the printed text have been ammended and noted. I have retained the accents, punctuation, and capitalization of the 1752 edition, except that the long 's' has been replaced by the short 's' and Arabic numerals with the character '§' appear as section numbers where Berkeley used Roman numerals. Berkeley's footnotes appear with the asterisk and (where necessary) dagger character as reference marks, and my additions to his notes appear in square brackets. My own notes are numbered consecutively throughout the text.

As in any translation, I have had to balance considerations of readability against those of accuracy. The result is a fairly literal translation which can still be read with relative ease. The case distinctions avaliable in Latin enable complex sentence structures which must be rendered by several sentences in English, but I have avoided the temptation to disregard Berkeley's sentence structure entirely and engage in a free translation. I have compared my efforts with those of A. A. Luce in the *Works* and G. N. Wright's 1843 translation in his edition of Berkeley's works. I found them helpful in certain difficult passages, but this translation is my own. I have retained the Latin title even for the English translation because the work is universally known in the literature as *De Motu* rather than *On Motion*.

5. BIBLIOGRAPHY

Aiton, Eric J. 1972. *The Vortex Theory of Planetary Motions*. London: Macdonald.

Alexander, H. G., ed. 1956. *The Leibniz-Clarke Correspondence*. Manchester: Manchester University Press.

Ardley, G. W. R. 1962. *Berkeley's Philosophy of Nature*. University of Auckland Bulletin No. 63, Philosophy Series No 3. Auckland, NZ: University of Auckland.

Aristotle. 1984. *The Complete Works of Aristotle: The Revised Oxford Translation*. Ed. Jonathan Barnes, 2 vols. Princeton, NJ: Princeton University Press.

Berkeley, George. 1948-57. *The Works of George Berkeley, Bishop of Cloyne*. Ed. A. A. Luce and T. E. Jessop. Edinburgh and London: Nelson.

Bertrand, Joseph. [1869] 1969. *L'Académie des sciences et les academiciens de 1666 a 1793*. Amsterdam: B. M. Israel.

Borelli, [Giovanni Alphonso]. 1667. *De Vi Percussionis Liber Io: Alphonsi Borelli*. Boglona: Ex typographia Iacobi Montii.

Brook, Richard J. 1973. *Berkeley's Philosophy of Science*. Archives Internationales D'Histoire Des Idees, no. 65. The Hague: Martinus Nijhoff.

Brykman, Geneviève. 1979. Le Cartesianisme dans le *De Motu*. *Revue internationale de philosophie* 33: 552-69.

Buchdahl, Gerd. 1969. *Metaphysics and the Phlosophy of Science; The Classical Origins: Descartes to Kant*. Cambridge, MA: MIT Press.

Burtt, E. A. 1950. *The Metaphysical Foundations of Modern Physical Science*. Rev. ed. London: Routledge & Kegan Paul.

Clagett, Marshall. 1959. *The Science of Mechanics in the Middle Ages*. University of Wisconsin Publications in Medieval Science. Madison, WI: University of Wisconsin Press.

Cohen, I. Bernard. 1980. *The Newtonian Revolution: With Illustrations of the Transformation of Scientific Ideas*. Cambridge and New York: Cambridge University Press.

Cohen, I. Bernard. 1985a. *The Birth of a New Physics*. Rev. ed. New York and London: Norton.

Cohen, I. Bernard. 1985b. *Revolution in Science*. Cambridge, MA: Harvard University Press.

Costabel, Pierre. 1967. Essai critique sur quelques concepts de la mécanique cartésienne. *Archives internationales d'histoire des sciences*. 20: 235-52.

Crousaz, [Jean-Pierre de]. 1728. *Essay sur le Mouvement; Où il es traité de sa Nature, de son Origine, de sa Communication, des Chocs des Corps qu'on suppose parfaitement Solides, du Plein & du Vuide, & de la Nature de la Reaction.* The Hague: Alberts and Vander Kloot.

Cudworth, Ralph. [1678] 1978. *The True Intellectual System of the Universe.* British Philosophers and Theologians of the Seventeenth and Eighteenth Centuries. New York: Garland.

Descartes, René. [1897-1909] 1964-1976. *Oeuvres de Descartes.* Ed. C. Adam and P. Tannery. 12 vols. Paris: Vrin/C.N.R.S.

Descartes, René. 1984. *The Philosophical Writings of Descartes.* Ed. and trans. John Cottingham, Robert Stoothoff, and Dugald Murdoch. 2 vols. Cambridge: Cambidge University Press.

Dijksterhuis, E. J. 1961. *The Mechanization of the World Picture.* Trans. C. Dikshoorn. Oxford: Oxford University Press.

Drake, Stillman and I. E. Drabkin, ed. and trans. 1969. *Mechanics in Sixteenth-Century Italy: Selections from Tartaglia, Benedetti, Guido Ubaldo, and Galileo.* University of Wisconsin Publications in Medieval Science. Madison, WI: University of Wisconsin Press.

Gabbey, Alan. 1980 Force and Inertia in the Seventeenth Century: Descartes and Newton. In Gaukroger 1980, 230-320. Harvester Readings in the History of Science and Philosophy. Sussex: Harvester Press; New York: Barnes and Noble.

Galilei, Galileo. 1957. *Discoveries and Opinions of Galileo.* Ed. and trans. Stillman Drake. New York: Anchor Doubleday.

Galilei, Galileo. 1967. *Dialogue Concerning the Two Chief World Systems - Ptolemaic & Copernican.* Ed. and trans. Stillman Drake, foreward by Albert Einstein. 2d ed. Berkeley, Los Angeles, and London: University of California Press.

Galilei, Galileo. 1974. *Two New Sciences; Including Centers of Gravity & Force of Percussion.* Ed. and trans. Stillman Drake. Madison, WI: University of Wisconsin Press.

Gaukroger, Steven, ed. 1980. *Descartes: Philosophy, Mathematics, and Physics.* Harvester Readings in the History of Science and Philosophy. Sussex: Harvester Press; New York: Barnes and Noble.

Grant, Edward. [1971] 1977. *Physical Science in the Middle Ages*. Cambridge History of Science Series. Cambridge: Cambridge University Press.

Guerlac, Henry. 1981. *Newton on the Continent*. Ithaca, NY: Cornell University Press.

Gueroult, Martial. 1934. *Dynamique et métaphysique Leibniziennes*. Paris: Vrin.

Gueroult, Martial. 1980. The Metaphysics of Force in Descartes. In Gaukroger 1980, 196-229.

Hall, A. Rupert. [1962] 1968. *The Scientific Revolution, 1500-1800: The Formation of the Modern Scientific Attitude*. 2d ed. Boston: Beacon Press.

Hall, A. Rupert. [1963] 1981. *From Galileo to Newton*. New York: Dover.

Hankins, Thomas L. 1965. Eighteenth-century Attempts to Resolve the *Vis Viva* Controversy. *Isis* 56: 281-97.

Hinrich, Gerard. 1950. The Logical Positivism of *De Motu*. *Review of Metaphysics* 3: 491-505.

Iltis, Carolyn. 1970. D'Alembert and the *Vis Viva* Controversy. *Studies in History and Philosophy of Science* 1: 135-44.

Iltis, Carolyn. 1971. Leibniz and the *Vis Viva* Controversy. *Isis* 62: 21-35.

Iltis, Carolyn. 1973a. The Decline of Cartesianism in Mechanics. *Isis* 64: 356-73.

Iltis, Carolyn. 1973b. The Leibnizian-Newtonian Debates: Natural Philosophy and Social Psychology. *British Journal for the History of Science* 6: 343-77.

Koyré, Alexandre. 1939. *Études Galiléennes*. Paris: Hermann.

Koyré, Alexandre. 1957. *From the Closed World to the Infinite Universe*. Baltimore, MD: Johns Hopkins University Press.

Koyré, Alexandre. 1965. *Newtonian Studies*. Cambridge, MA: Harvard University Press.

Laudan, L. L. 1968. The *Vis Viva* Controversy: A Post-Mortem. *Isis* 59: 131-43.

Leibniz, G. W. [1849-55] 1962. *G. W. Leibniz Mathematische Schriften*. Ed. C. I. Gerhardt. 7 vols. Hildesheim: Olms.

Leibniz, G. W. 1989. *G. W. Leibniz: Philosophical Essays*. Ed. and trans. Roger Ariew and Daniel Garber. Indianapolis, IN and Cambridge, MA: Hackett Publishing Company.

McMullin, Ernan, ed. 1967. *Galileo: Man of Science*. New York and London: Basic Books.

Mirarchi, Lawrence A. 1977a. Force and Absolute Motion in Berkeley's Philosophy of Physics. *Journal of the History of Ideas* 38: 705-13.

Mirarchi, Lawrence A. 1977b. A Rejoinder to Bruce Silver's Reply. *Journal of the History of Ideas* 38: 716-18.

Mirarchi, Lawrence A. 1982. Dynamical Implications of Berkeley's Doctrine of Heterogeneity: A Note on the Language Model of Nature. In *Berkeley: Critical and Interpretive Essays*, ed. C. M. Turbayne, 247-60. Minneapolis, MN: University of Minnesota Press.

Moscovici, Serge. 1968. Torricelli's *Lezioni Accademiche* and Galileo's theory of percussion. In McMullin 1968, 432-48.

Myhill, John. 1957. Berkeley's *De Motu* – An Anticipation of Mach. In *George Berkeley: Lectures delivered before the Philosophical Union of the University of California in honor of the two hundredth anniversary of the death of George Berkeley, Bishop of Cloyne*. Ed. Steven C. Pepper, Karl Aschenbrenner and Benson Mates, 65-88. University of California Publications in Philosophy no. 29. Berkeley and Los Angeles: University of California Press.

Newton, Isaac. [1727] 1934. *Sir Isaac Newton's Mathematical Principles of Natural Philosophy and his System of the World*. Trans. Andrew Motte, ed. and rev. Florian Cajori. 2 vols. Berkeley, Los Angeles, and London: University of California Press.

Newton, Isaac. [1726] 1972. *Philosophiæ naturalis principia mathematica*. Ed. Alexandre Koyré, I. Bernard Cohen, and Anne Whitman. Cambridge, MA: Harvard University Press.

Newton-Smith, W. H. 1985. Berkeley's Philosophy of Science. In *Essays on Berkeley: A Tercentennial Celebration*, ed. John Foster and Howard Robinson 149-161. Oxford: Clarendon Press.

Papineau, David. 1977. The *Vis Viva* Controversy. *Studies in History and Philosophy of Science* 8: 111-42.

Popper, Karl. 1953. A Note on Berkeley as a Precursor of Mach and Einstein. *British Journal for the Philosophy of Science* 4: 26-36.

Raphson, Joseph. 1697. *Analysis Æquationum universalis seu Ad æqua-tiones algebraicas resolvendas methodus generalis, et expedita, ex nova infinitarum serierum methodo, deducta ac demonstrata. Editio secunda cum appendice. Cui annexum est, De Spatio Reali, seu Ente Infinito Conamen Mathematico-Metaphysicum*. London: Churchil.

Scheurer, P. B. and G. Debrock, ed. 1988. *Newton's Scientific and Philosophical Legacy*. Dordrecht, Boston, and London: Kluwer Academic Publishers.

Silver, Bruce. 1973. Berkeley and the Principle of Inertia. *Journal of the History of Ideas* 34: 599-608.

Silver, Bruce. 1977. Reply to Professor Mirarchi on Force and Absolute Motion. *Journal of the History of Ideas* 38: 714-715.

Stock, Joseph. [1776] 1989. *An Account of the Life of George Berkeley, D. D. Late Bishop of Cloyne in Ireland. With Notes, Containing Strictures Upon his Works*. Reprint in David Berman, ed. 1989. *George Berkeley: Eighteenth Century Responses*. 2 vols. New York: Garland, 1: 5-85.

Suchting, W. A. 1967. Berkeley's Criticism of Newton on Space and Time. *Isis* 58: 186-197.

Torricelli, Evangelista. 1715. *Lezioni Accademiche D'Evangelista Torricelli*. Florence: Jacopo Guiducci.

Westfall, Richard S. 1971. *Force in Newton's Physics*. London: Macdonald & Co.; New York: American Elsevier.

Westfall, Richard S. [1971] 1978. *The Construction of Modern Science: Mechanisms and Mechanics*. Cambridge History of Science Series. Cambridge: Cambridge University Press.

Westfall, Richard S. 1980. *Never at Rest: A Biography of Isaac Newton*. Cambridge: Cambridge University Press.

Whitrow, G. J. 1953a. Berkeley's Philosophy of Motion. *British Journal for the Philosophy of Science*. 4: 37-45.

Whitrow, G. J. 1953b. Berkeley's Critique of the Newtonian Analysis of Motion. *Hermathena* 82: 90-112.

Wright, Rev. G. N., ed. 1843. *The Works of George Berkeley, D. D.* 2 vols. London: Tegg.

D E

M O T U;

S I V E D E

Motus Principio & Natura,

E T D E

Cauſa Communicationis Motuum.

Diatriba primùm Typis mandata,

L O N D I N I. A. D. M DCC XXI.

DE MOTU;

SIVE

De Motus Principio & Natura, et de Causa Communicationis Motuum

§1. Ad veritatem inveniendam præcipuum est cavisse ne voces malè intellectæ nobis officiant: quod omnes fere monent philosophi, pauci observant. Quanquam id quidem haud adeo difficile videtur, in rebus præsertim Physicis tractandis, ubi locum habent sensus, experientia, & ratiocinium geometricum. Seposito igitur, quantum licet, omni præjudicio, tam à loquendi consuetudine, quam à philosophorum auctoritate nato, ipsa rerum natura diligenter inspicienda. Neque enim cujusquam auctoritatem usque adeo valere oportet, ut verba ejus & voces in pretio sint, dummodo nihil clari & certi iis subesse comperiatur.

§2. Motus contemplatio mirè torsit veterum philosophorum mentes, unde natæ sunt variæ opiniones supra modum difficiles, ne dicam absurdæ, quæ quum jam fere in desuetudinem abierint, haud merentur ut iis discutiendis nimio studio immoremur. Apud recentiores autem & saniores hujus ævi Philosophos, ubi de motu agitur, vocabula haud pauca abstractæ nimium & obscuræ significationis occurrunt, cujusmodi sunt *solicitatio gravitatis, conatus, vires mortuæ,* &c. quæ scriptis alioqui doctissimis tenebras offundunt, senteniisque, non minus à vero quam à sensu hominum communi abhorrentibus ortum præbent. Hæc vero necesse est ut, veritatis gratia, non alios refellendi studio, accuratè discutiantur.

§3. Solicitatio & nisus sive conatus rebus solummodo animatis revera competunt. Cum aliis rebus tribuuntur, sensu metaphorico accipiantur necesse est. A metaphoris autem abstinendum philosopho. Porro seclusâ omni tam animæ affectione quam corporis motione, nihil clari ac distincti iis vocibus significari cuilibet constabit, qui modò rem seriò perpenderit.

§4. Quamdiu corpora gravia à nobis sustinentur, sentimus in nobismet ipsis nisum, fatigationem, & molestiam. Percipimus etiam in gravibus cadentibus motum acceleratum versus centrum telluris: ope sensum præterea nihil. Ratione tamen colligitur causam esse aliquam vel principium horum phænomenôn, illud autem *gravitas* vulgò nuncupatur. Quoniam verò causa descensus gravium cæca sit & incognita: gravitas ea acceptione propriè dici nequit qualitas sensibilis: est igitur qualitas occulta. Sed vix, & ne vix quidem, concipere licet quid sit qualitas occulta, aut qua ratione qualitas ulla agere aut operari quidquam possit. Melius itaque foret, si, missa qualitate occulta, homines attenderent solummodo ad effectus sensibiles, vocibusque abstractis, (quantumvis illæ ad disserendum utiles sint) in meditatione omissis, mens in particularibus & concretis, hoc est in ipsis rebus, defigeretur.

§5. *Vis* similiter corporibus tribuitur; usurpatur autem vocabulum illud, tamquam significaret qualitatem cognitam, distinctamque tam à motu, figura, omnique alia re sensibili, quam ab omni animalis affectione, id vero nihil aliud esse quàm qualitatem occultam rem acriùs rimanti constabit. Nisus animalis & motus corporeus vulgo spectantur tanquam symptomata & mensuræ hujus qualitatis occultæ.

§6. Patet igitur gravitatem aut vim frustra poni pro principio motus: nunquid enim principium illud clarius cognosci potest ex eo quod dicatur qualitas occulta? Quod ipsum occultum est nihil explicat. Ut omittamus causam agentem incognitam rectius dici posse substantiam quam qualitatem. Porro, *vis, gravitas, & istiusmodi* voces sæpius, nec ineptè, in concreto usurpantur, ita ut connotent corpus motum, difficultatem resistendi, *&c.* Ubi vero à Philosophis adhibentur ad significandas naturas quasdam ab hisce omnibus præcisas & abstractas, quæ nec sensibus subjiciuntur nec ulla mentis vi intelligi nec imaginatione effingi possunt, tum demùm erores & confusionem pariunt.

§7. Multos autem in errorem ducit, quod voces generales & abstractas in disserendo utiles esse videant, nec tamen earum vim satis capiant. Partim vero à consuetudine vulgari inventæ sunt illæ ad sermonem abbreviandum, partim, à Philosophis ad docendum excogitatæ: non, quod ad naturas rerum accomodatæ sint, quæ quidem singulares, & concretæ existunt, sed quod idoneæ ad tradendas disciplinas, propterea quod faciant notiones vel saltem propositiones universales.

§8. Vim corpoream esse aliquid conceptu facile plerumque existimamus: ii tamen qui rem accuratiùs inspexerunt in diversa sunt opinione, uti apparet ex mira verborum obscuritate qua laborant, ubi illam explicare conantur. Torricellius ait vim & impetum esse res quasdam abstractas subtilesque, & quintessentias quæ includuntur in substantia corporea, tanquam in vase magico Circes.* Leibnitius item in natura vis explicanda hæc habet. *Vis activa, primitiva, quæ est* ἐντελέχεια'η πρώτη, *animæ vel formæ substantiali respondet. vid. Acta erudit. Lips.* Usque

---

* La materia altro non e che un vaso di Circe incantato, il quale serve per ricettacolo della forza & de momenti dell' impeto. La forza & l'impeti sono astratti tanto sottili, sono quintessenze tanto spiritose, che in altre ampolle non se possono racchiudere, fuor che nell' intima corpulenza de solidi naturali. Vid. *Lezioni Academiche.*

adeo necesse est ut vel summi viri quamdiu abstractionibus indulgent, voces nulla certa significatione præditas & meras scholasticorum umbras sectentur. Alia ex neotericorum scriptis, nec pauca quidem ea, producere liceret, quibus abunde constaret, metaphysicas abstractiones non usquequaque cessisse mechanicæ & experimentis, sed negotium inane philosophis etiamnum facessere.

§9. Ex illo fonte derivantur varia absurda cujus generis est illud, *vim percussionis utcunque exiguæ esse infinitè magnam.* Quod sane supponit, gravitatem esse qualitatem quandam realem ab aliis omnibus diversam: & gravitationem esse quasi actum hujus qualitatis à motu realiter distinctum; minima autem percussio producit effectum majorem quam maxima gravitatio sine motu. Illa scilicet motum aliquem edit, hæc nullum. Unde sequitur, vim percussionis ratione infinita excedere vim gravitationis, hoc est esse infinitè magnam. Videantur experimenta Galilæi & quæ de infinita[1] vi percussionis scripserunt Torricellius, Borellus, & alii.

§10. Veruntamen fatendum est vim nullam per se immediate sentiri, neque aliter quam per effectum cognosci & mensurari; sed vis mortuæ seu gravitationis simplicis, in corpore quiescente subjecto nulla facta mutatione, effectus nullus est. Percussionis autem, effectus aliquis. Quoniam ergo vires sunt effectibus proportionales: concludere licet vim mortuam esse nullam: neque tamen propterea vim percussionis esse infinitam: non enim oportet quantitatem ullam positivam habere pro infinita, propterea quod ratione infinita superet quantitatem nullam sive nihil.

§11. Vis gravitationis à momento secerni nequit, momentum autem sine celeritate nullum est, quum sit moles in celeritatem ducta, porro

---

[1] Reading 'infinita' for 'definita' in accordance with the first edition.

celeritas sine motu intelligi non potest, ergo nec vis gravitationis. Deinde, vis nulla nisi per actionem innotescit & per eandem mensuratur, actionem autem corporis à motu præscindere non possumus, ergo, quamdiu corpus grave plumbi subjecti vel chordæ figuram mutat, tamdiu movetur: ubi vero quiescit, nihil agit, vel, quod idem est, agere prohibetur. Breviter, voces istæ *vis mortua* & *gravitatio*, etsi per abstractionem metaphysicam aliquid significare supponuntur diversum à movente, moto, motu & quiete, reverâ tamen id totum nihil est.

§12. Siquis diceret pondus appensum vel impositum agere in chordam, quoniam impedit quominus se restituat vi elastica: dico, pari ratione corpus quodvis inferum agere in superius incumbens, quoniam illud descendere prohibet: dici vero non potest actio corporis, quod prohibeat aliud corpus existere in eo loco quem occupat.

§13. Pressionem corporis gravitantis quandoque sentimus. Verum sensio ista molesta oritur ex motu corporis istius gravis fibris nervisque nostri corporis communicatio, & eorundem situm immutante, adeoque percussioni accepta referri debet. In hisce rebus multis & gravibus præjudiciis laboramus, sed illa acri atque iteratâ meditatione subigenda sunt, vel potius penitùs averruncanda.

§14. Quo probetur, quantitatem ullam esse infinitam, ostendi oportet partem aliquam finitam homogeneam in eâ infinities contineri. Sed vis mortua se habet ad vim percussionis non ut pars ad totum, sed ut punctum ad lineam, juxta ipsos vis infinitæ percussionis auctores. Multa in hanc rem adjicere liceret sed vereor ne prolixus sim.

§15. Ex principiis præmissis lites insignes solvi possunt, quæ viros doctos multum exercuerunt. Hujus rei exemplum sit controversia illa de proportione virium. Una pars dum concedit, momenta, motus, impetus, data mole, esse simpliciter ut velocitates, affirmat vires esse ut quadrata velocitatum. Hanc autem sententiam supponere, vim corporis distingui à momento, motu, & impetu, eaque suppositione sublata corruere, nemo non videt.

§16. Quo clarius adhuc appareat, confusionem quandam miram per abstractiones metaphysicas in doctrinam de motu introductam esse, videamus quantum intersit inter notiones virorum celebrium de vi & impetu. Leibnitius impetum cum motu confundit. Juxta Newtonum impetus revera idem est cum vi interiæ. Borellus asserit impetum non aliud esse quam gradum velocitatis. Alii impetum & conatum inter se differre, alli non differre volunt. Plerique vim motricem motui proportionalem intelligunt, nonnulli aliam aliquam vim præter motricem, & diversimodè mensurandam, utpote per quadrata velocitatum in moles, intelligere præ se ferunt. Sed infinitum esset hæc prosequi.

§17. *Vis, gravitas, attractio,* & hujusmodi voces utiles sunt ad ratiocinia, & computationes de motu & corporibus motis: sed non ad intelligendam simplicem ipsius motus naturam, vel ad qualitates totidem distinctas designandas. Attractionem certe quod attinet, patet illam ab Newtono adhiberi, non tanquam qualitatem veram & physicam, sed solummodo ut hypothesin mathematicam. Quin & Leibnitius, nisum elementarem seu solicitationem ab impetu distinguens, fatetur illa entia non re ipsa inveniri in rerum natura, sed abstractione facienda esse.

§18. Similis ratio est compositionis & resolutionis [2] virium quarumcunque directarum in quascunque obliquas, per diagonalem & latera parallelogrammi. Hæc mechanicæ & computationi inserviunt: sed aliud est computationi & demonstrationibus mathematicis inservire, aliud, rerum naturam exhibere.

---

[2] Reading 'resolutionis' for 'resolutiones', following the first edition.

§19. Ex recentioribus multi sunt in eâ opinione, ut putent motum neque destrui nec de novo gigni, sed eandem semper motus quantitatem permanere. Aristoteles etiam dubium illud olim proposuit, utrum motus factus sit & corruptus, an vero ab æterno? phys. 1. 8. Quod vero motus sensibilis pereat, patet sensibus, illi autem eundem impetum, nisum, aut summam virium eandem manere velle videntur. Unde affirmat Borellus, vim in percussione, non imminui sed expandi, impetus etiam contrarios suscipi & retinere in eodem corpore. Item Leibnitius nisum ubique & semper esse in materia, &, ubi non patet sensibus, ratione intelligi contendit. Hæc autem nimis abstracta esse & obscura, ejusdemque ferè generis cum formis substantialibus & Entelechiis, fatendum.

§20. Quotquot ad explicandam motus causam atque originem vel principio Hylarchico, vel naturæ indigentiâ, vel appetitu, aut denique instinctu naturali utuntur, dixisse aliquid potius quam cogitâsse censendi sunt. Neque ab hisce multùm absunt qui supposuerint* *partes terræ esse se moventes, aut etiam spiritus iis implantatos ad instar formæ*, ut assignent causam accelerationis gravium cadentium. Aut qui dixerit† *in corpore præter solidam extensionem debere etiam poni aliquid unde virium consideratio oriatur.* Siquidem hi omnes vel nihil particulare & determinatum enuntiant: vel, si quid sit, tam difficile erit illud explicare, quam id ipsum cujus explicandi causâ adducitur.

§21 Frustra ad naturam illustrandam adhibentur ea quæ nec sensibus patent, nec ratione intelligi possunt. Videndum ergo quid sensus, quid experientia, quid demùm ratio iis innixa suadeat. Duo sunt summa rerum genera, corpus & anima. Rem extensam, solidam, mobilem, figuratam, aliisque qualitatibus quæ sensibus occurrunt præditam, ope

---

* Borellus
† Leibnitius

sensuum, rem vero sentientem, percipientem, intelligentem, conscientiâ quâdam internâ cognovimus. Porro, res istas planè inter se diversas esse, longèque heterogeneas, cernimus. Loquor autem de rebus cognitis, de incognitis enim disserere nil juvat.

§22 Totum id quod novimus, cui nomen *corpus* indidimus, nihil in se continet quod motus principium seu causa efficiens esse possit; etenim impenetrabilitas, extensio, figura nullam includunt vel connotant potentiam producendi motum: quinimò è contrario non modo illas verum etiam alias, quotquot sint, corporis qualitates sigillatim percurrentes, videbimus omnes esse revera passivas, nihilque iis activum inesse, quod ullo modo intelligi possit tanquam fons & principium motus. Gravitatem quod attinet, voce illa nihil cognitum & ab ipso effectu sensibili, cujus causa quæritur, diversum significari jam ante ostendimus. Et sanè quando corpus grave dicimus nihil aliud intelligimus, nisi quod feratur deorsum, de causâ hujus effectus sensibilis nihil omnino cogitantes.

§23 De corpore itaque audacter pronunciare licet, utpote de re comperta, quod non sit principium motûs. Quod si quisquam, præter solidam extensionem ejusque modificationes, vocem *corpus* qualitatem etiam occultam, virtutem, formam, essentiam complecti sua significatione contendat; licet quidem illi inutili negotio sine ideis disputare, & nominibus nihil distinctè exprimentibus abuti. Cæterùm sanior philosophandi ratio videtur ab notionibus abstractis & generalibus (si modo notiones dici debent quæ intelligi nequeunt) quantum fieri potest abstinuisse.

§24 Quicquid continetur in idea corporis novimus: quod vero novimus in corpore id non esse principium motûs constat. Qui præterea aliquid incognitum in corpore, cujus ideam nullam habent, comminiscuntur, quod motûs principium dicant: ii revera nihil aliud quam principium motus esse incognitum dicunt. Sed hujusmodi subtilitatibus diutiùs immorari piget.

§25 Præter res corporeas alterum est genus rerum cogitantium, in iis autem potentiam inesse corpora movendi, propria experientia didicimus, quandoquidem anima nostra pro libitu possit ciere & sistere membrorum motus, quacunque tandem ratione id fiat. Hoc certè constat, corpora moveri ad nutum animæ, eamque proinde haud ineptè dici posse principium motus; particulare quidem & subordinatum, quodque ipsum dependeat à primo & universali principio.

§26 Corpora gravia feruntur deorsum, etsi nullo impulsu apparente agitata, non tamen existimandum propterea in iis contineri principium motus: cujus rei hanc rationem assignat Aristoteles, *gravia & levia*, inquit, *non moventur à seipsis, id enim vitale esset, & se sistere possent.* Gravia omnia unâ eâdemque certâ & constanti lege centrum telluris petunt, neque in ipsis animadvertitur principium vel facultas ulla motum istum sistendi, minuendi vel, nisi pro rata proportione, augendi, aut denique ullo modo immutandi: habent adeò se passivè. Porro idem, strictè & accuratè loquendo, dicendum de corporibus percussivis. Corpora ista quamdiu moventur, ut & in ipso percussionis momento, se gerunt passivè, perinde scilicet atque cum quiescunt. Corpus iners tam agit quam corpus motum, si res ad verum exigatur: id quod agnoscit Newtonus, ubi ait, vim interiæ esse eandem cum impetu. Corpus autem iners & quietum nihil agit, ergo nec motum.

§27 Revera corpus æquè preseverat in utrovis statu, vel motûs vel quietis. Ista vero perseverantia non magis dicenda est actio corporis, quam existentia ejusdem actio diceretur. Perseverantia nihil aliud est quam continuatio in eodem modo existendi, quæ propriè dici actio non potest. Cæterùm resistentiam, quam experimur in sistendo corpore moto, ejus actionem esse fingimus vana specie delusi. Revera enim ista resistentia quam sentimus, passio est in nobis, neque arguit corpus agere, sed nos pati: constat utique nos idem passuros fuisse, sive corpus illud à se moveatur, sive ab alio principio impellatur.

§28 Actio & reactio dicuntur esse in corporibus: nec incommodè ad demonstrationes mechanicas. Sed cavendum, ne propterea supponamus virtutem aliquam realem quæ motûs causa, sive principium sit, esse in iis. Etenim voces illæ eodem modo intelligendæ sunt ac vox *attractio*, & quemadmodum hæc est hypothesis solummodo mathematica non autem qualitas physica; idem etiam de illis intelligi debet, & ob eandem rationem. Nam sicut veritas & usus theorematum de mutua corporum attractione in philosophia mechanica stabiles manent, utpote unicè fundati in motu corporum, sive motus iste causari supponatur per actionem corporum se mutuo attrahentium, sive per actionem agentis alicujus à corporibus diversi impellentis & moderantis corpora; pari ratione, quæcunque tradita sunt de regulis & legibus motuum, simul ac theoremata inde deducta, manent inconcussa, dummodo concedantur effectus sensibiles, & ratiocinia iis innixa; sive supponamus actionem ipsam, aut vim horum effectuum causatricem, esse in corpore, sive in agente incoproreo.

§29 Auferantur ex idea corporis extensio, soliditas, figura, remanebit nihil. Sed qualitates istæ sunt ad motum indifferentes, nec in se quidquam habent, quod motus principium dici possit. Hoc ex ipsis ideis nostris perspicuum est. Si igitur voce *corpus* significatur, id quod concipimus: planè constat inde non peti posse principium motus: pars scilicet nulla aut attributum illius causa efficiens vera est, quæ motum producat. Vocem autem proferre, & nihil concipere, id demùm indignum esset philosopho.

§30 Datur res cogitans activa quam principium motûs esse in nobis experimur. Hanc *animam, mentem, spiritum* dicimus; datur etiam res extensa, iners, impenetrabilis, mobilis, quæ à priori toto cœlo differt, novumque genus constituit. Quantum intersit inter res cogitantes & extensas, primus omnium deprehendens Anaxagoras vir longè sapientissimus, asserebat mentem nihil habere cum corporibus commune, id quod constat ex primo libro Aristotelis de anima. Ex neotericis idem optimè animadvertit Cartesius. Ab eo alii rem satis claram vocibus obscuris impeditam ac difficilem reddiderunt.

§31 Ex dictis manifestum est eos qui vim activam, actionem, motus principium, in corporibus revera inesse affirmant, sententiam nulla experientia fundatam amplecti, eamque terminis obscuris & generalibus adstruere, nec quid sibi velint satis intelligere. E contrario, qui mentem esse principium motus volunt, sententiam propria experientia munitam proferunt,[3] hominumque omni ævo doctissimorum suffragiis comprobatam.

§32 Primus Anaxagoras τὸν νοῦν introduxit, qui motum inerti materiæ imprimeret, quam quidem sententiam probat etiam Aristoteles pluribusque confirmat, apertè pronuncians primum movens esse immobile, indivisibile, & nullam habens magnitudinem. Dicere autem, omne motivum esse mobile, rectè animadvertit idem esse ac siquis diceret, omne ædificativum esse ædificabile, physic. 1. 8. Plato insuper in Timæo tradit machinam hanc corpoream, seu mundum visibilem agitari & animari à mente, quæ sensum omnem fugiat. Quinetiam hodie, philosophi Cartesiani principium motuum naturalium Deum agnoscunt. Et Newtonus passim nec obscurè innuit, non solummodo motum ab initio à numine profectum esse, verum adhuc systema mundanum ab eodem actu moveri. Hoc sacris literis consonum est: hoc scholasticorum

---

[3] Reading 'proferunt' for 'proferent' with the first edition.

calculo comprobatur. Nam etsi peripatetici naturam tradant esse principium motûs & quietis, interpretantur tamen naturam naturantem esse Deum. Intelligunt nimirum corpora omnia systematis hujusce mundandi à mente præpotenti, juxta certam & constantem rationem moveri.

§33 Cæterùm qui principium vitale corporibus tribuunt, obscurum aliquid & rebus parùm conveniens fingunt. Quid enim aliud est vitali principio præditum esse quam vivere? aut vivere quam se movere, sistere, & statum suum mutare? Philosophi autem hujus saeculi doctissimi pro principio indubitato ponunt, omne corpus perservare in statu suo, vel quietis vel motûs uniformis in directum, nisi quatenus aliunde cogitur statum illum mutare; è contrario, in anima sentimus esse facultatem tam statum suum quam aliarum rerum mutandi; id quod propriè dicitur vitale, animamque à corporibus longe discriminat.

§34 Motum & quietem in corporibus recentiores considerant velut duos status existendi, in quorum utrovis corpus omne sua natura iners permaneret, nulla vi externa urgente. Unde colligere licet, eandem esse causam motûs & quietis, quæ est existentiæ corporum. Neque enim quærenda videtur alia causa existentiæ corporis successivæ in diversis partibus spatii, quam illa unde derivatur existentia ejusdem corporis successiva in diversis partibus temporis. De Deo autem optimo maximo rerum omnium conditore & conservatore tractare: & qua ratione res cunctæ à summo & vero ente pendeant demonstrare, quamvis pars sit scientiæ humanæ præcellentissima, spectat tamen potius ad philosophiam primam seu metaphysicam & theologiam, quam ad philosophiam naturalem, quæ hodie fere omnis continetur in experimentis & mechanicâ. Itaque cognitionem de Deo vel supponit philosophia naturalis, vel mutuatur ab aliqua scientia superiori. Quanquam verissimum sit, naturæ investigationem scientiis altioribus argumenta egregia ad sapientiam, bonitatem & potentiam Dei illustrandam & probandam undequaque subministrare.

§35 Quod hæc minus intelligantur, in causa est, cur nonnulli immertio repudient physicæ principia mathematica, eo scilicet nomine quod illa causas rerum efficientes non assignant. Quum tamen revera ad physicam aut mechanicam spectet regulas solummodo, non causas efficientes, impulsionum attractionumve &, ut verbo dicam, motuum leges tradere: ex iis vero positis phænomenôn particularium solutionem, non autem, causam efficientem assignare.

§36 Multum intererit considerasse quid propriè sit principium, & quo sensu intelligenda sit vox illa apud philosophos. Causa quidem vera efficiens, & conservatrix rerum omnium jure optimo appellatur fons & principium earundem. Principia vero philosophiæ experimentalis propriè dicenda sunt fundamenta, quibus illa innititur, seu fontes unde derivatur, (non dico existentia, sed) cognitio rerum corporearum, sensus utique & experientia. Similiter, in philosophia mechanica, principia dicenda sunt, in quibus fundatur & continetur universa disciplina, leges, illæ motum primariæ, quæ experimentis comprobatæ, ratiocinio etiam excultæ sunt & redditæ universales. Hæ motuum leges commodè dicuntur principia, quoniam ab iis tam theoremata mechanica generalia quam particulares τῶν Φαινομένων explicationes derivantur.

§37 Tum nimirum dici potest quidpiam explicari mechanicè, cum reducitur ad ista principia simplicissima & universalissima, & per accuratum ratiocinium, cum iis consentaneum & connexum esse ostenditur. Nam, inventis semel naturæ legibus, deinceps monstrandum est philosopho, ex constanti harum legum observatione, hoc est, ex iis principiis phænomena quodvis necessario consequi: id quod est phænomena explicare & solvere, causamque, id est rationem cur fiant, assignare.

§38 Mens humana gaudet scientiam suam extendere & dilatare. Ad hoc autem notiones & propositiones generales efformandæ sunt, in quibus quodam modo continentur propositiones & cognitiones particulares, quæ tum demùm intelligi creduntur.[4] Hoc geometris notissimum est. In mechanica etiam præmittuntur notiones, hoc est definitiones, et enunciationes de motu primæ & generales, ex quibus postmodùm methodo mathematica conclusiones magis remotæ, & minus generales colliguntur. Et sicut per applicationem theorematum geometricorum, corporum particularium magnitudines mensurantur; ita etiam per applicationem theorematum mechanices universalium, systematis mundani parium quarumvis motus, & phænomena inde pendentia innotescunt & determinantur: ad quem scopum uniquè collineandum physico.

§39 Et quemadmodum geometræ disciplinæ causa, multa comminiscuntur, quæ nec ipsi describere possunt, nec in rerum natura invenire: simili prorsus ratione mechanicus voces quasdam abstractas & generales adhibet, fingitque in corporibus vim, actionem, attractionem, solicitationem, &c. quæ ad theorias & enunciationes, ut & computationes de motu apprime utiles sunt, etiamsi in ipsâ rerum veritate & corporibus actu existentibus frustra quærerentur, non minus quàm quæ à geometris per abstractionem mathematicam finguntur.

§40 Revera, ope sensum nihil nisi effectus seu qualitates sensibiles, & res corporeas omnino passivas, sive in motu sint sive in quiete, percipimus: ratioque & experientia activum nihil præter mentem aut animam esse suadet. Quid quid ultra fingitur, id ejusdem generis esse cum

---

[4] The 1721 edition adds "cum ex primis illis continuo nexu deducuntur."

aliis hypothesibus & abstractionibus mathematicis existimandum; quod penitus animo infigere oportet. Hoc ni fiat, facilè in obscuram scholasticorum subtilitatem, quæ per tot sæcula, tanquam dira quædam pestis, philosophiam corrupit, relabi possumus.

§41 Principia mechanica legesque motuum aut naturæ universales, sæculo ultimo feliciter inventæ, & subsidio geometriæ tractatæ & applicatæ, miram lucem in philosophiam intulerunt. Principia vero metaphysica causæque reales efficientes motus & existentiæ corporum attributorumve corporeorum nullo modo ad mechanicam aut experimenta pertinent, neque eis lucem dare possunt, nisi quatenus, velut præcognita inserviant ad limites physicæ præfiniendos, eaque ratione ad tollendas difficultates quæstionesque peregrinas.

§42 Qui à spiritibus motus principium petunt, ii vel rem corpoream vel incorpoream voce *spiritus* intelligunt: si rem corpoream, quantumvis tenuem, tamen redit difficultas: si incorpoream, quantiumvis id verum sit, attamen ad physicam non propriè pertinet. Quod si quis philosophiam naturalem ultra limites experimentorum & mechanicæ extenderit, ita ut rerum etiam incoprorearum, & inextensarum cognitionem complectatur: latior quidem illa vocis acceptio tractationem de anima, mente, seu principio vitali admittit. Cæterùm commodius erit, juxta usum jam ferè receptum, ita distinguere inter scientias, ut singulæ propriis circumscribantur cancellis, & philosophus naturalis totus sit in experimentis, legibusque motuum, & principiis mechanicis, indeque depromptis ratiociniis; quidquid autem de aliis rebus protulerit id superiori alicui scientiæ acceptum referat. Etenim ex cognitis naturæ legibus pulcherrimæ theoriæ, praxes etiam mechanicæ ad vitam utiles consequuntur. Ex cognitione autem ipsius naturæ auctoris considerationes, longe præstantissimæ quidem illæ, sed, metaphysicæ, theologicæ, morales oriuntur.

§43 De principiis hactenus: nunc dicendum de natura motus, atque is quidem, cum sensibus clare percipiatur non tam natura sua, quam doctis philosophorum commentis obscuratus est. Motus nunquam in sensus nostros incurrit sine mole corporea, spatio, & tempore. Sunt tamen qui motum, tanquam ideam quandam simplicem & abstractam, atque ab omnibus aliis rebus sejunctam, contemplari student. Verùm idea illa tenuissima & subtilissima intellectûs aciem eludit: id quod quilibet secum meditando experiri potest. Hinc nascuntur magnæ difficultates de natura motus, & definitiones, ipsa re quam illustrare debent, longe obscuriores. Hujusmodi sunt definitiones illæ Aristotelis & Scholasticorum, qui motum dicunt esse actum *mobilis, quatenus est mobile, vel actum entis in potentia quatenus in potentia.* Hujusmodi etiam est illud, viri inter recentiores celebris, qui asserit *nihil in motu esse reale præter momentaneum illud quod in vi ad mutationem nitente constitui debet.* Porro, constat, horum & similium definitionum auctores in animo habuisse abstractam motus naturam, seclusa omni temporis & spatii consideratione, explicare, sed qua ratione abstracta illa motus quintessentia (ut ita dicam) intelligi possit non video.

§44 Neque hoc contenti, ulterius pergunt partesque ipsius motus à se invicem dividunt & secernunt, quarum ideas distinctas, tanquam entium revera distinctorum, efformare conantur. Etenim sunt qui motionem à motu distinguant, illam velut instantaneum motus elementum spectantes. Velocitatem insuper, conatum, vim, impetum totidem res essentia diversas esse volunt, quarum quæque per propriam atque ab aliis omnibus segregatam & abstractam ideam intellectui objiciatur. Sed in hisce rebus discutiendis, stantibus iis quae supra disseruimus, non est cur diutius immoremur.

§45 Multi etiam per *transitum* motum definiunt, obliti scilicet transitum ipsum sine motu intelligi non posse, & per motum definiri oportere. Verissimum adeo est definitiones, sicut nonnullis rebus lucem, ita vicissim aliis tenebras afferre. Et profecto, quascumque res sensu percipimus, eas clariores aut notiores definiendo efficere vix quisquam potuerit. Cujus rei vana spe allecti res faciles difficillimas reddiderunt philosophi, mentesque suas difficultatibus, quas ut plurimum ipsi peperissent, implicavere. Ex hocce definiendi, simulac abstrahendi studio, multæ, tam de motu, quam de aliis rebus natæ subtilissimae quæstiones, eædemque nullius utilitatis, hominum ingenia frustra torserunt, adeo ut Aristoteles ultro & sæpius fateatur motum esse *actum quendam cognitu difficilem,* & nonnulli ex veteribus usque eo nugis exercitati deveniebant, ut motum omnino esse negarent.

§46 Sed hujusmodi minutiis distineri piget. Satis sit fontes solutionum indicasse: ad quos etiam illud adjungere libet: quod ea quæ de infinita divisione temporis & spatii in mathesi traduntur, ob congenitam rerum naturam paradoxa & theorias spinosas (quales sunt illæ omnes in quibus agitur de infinito) in speculationes de motu intulerunt. Quidquid autem hujus generis sit, id omne motus commune habet cum spatio & tempore, vel potius ad ea refert acceptum.

§47 Et quemadmodum, ex una parte nimia abstractio seu divisio rerum verè inseparabilium, ita, ab altera parte, compositio seu potius confusio rerum diversissimarum motus naturam perplexam reddidit. Usitatum enim est motum cum causa motus efficiente confundere. Unde accidit ut motus sit quasi biformis, unam faciem sensibus obviam, alteram caliginosa nocte obvolutam habens. Inde obscuritas & confusio,

& varia de motu paradoxa originem trahunt, dum effectui perperam tribuitur id quod revera causæ solummodo competit.

§48 Hinc oritur opinio illa, eandem semper motus quantitatem conservari; quod, nisi intelligatur de vi & potentia causæ, sive causa illa dicatur natura, sive *νοῦς*, vel quodcunque tandem agens sit, falsum esse cuivis facile constabit. Aristoteles quidem l.8. physicorum, ubi quærit *utrum motus factus sit & corruptus, an vero ab æterno tanquam vita immortalis insit rebus omnibus,* vitale principium potius, quam effectum externum, sive mutationem loci intellexisse videtur.

§49 Hinc etiam est, quod multi suspicantur motum non esse meram passionem in corporibus. Quod si intelligamus id quod, in motu corporis, sensibus objicitur, quin omnino passivum sit nemo dubitare potest. Ecquid enim in se habet successiva corporis existentia in diversis locis, quod actionem referat, aut aliud sit quam nudus & iners effectus?

§50 Peripatetici, qui dicunt motum esse actum unum utriusque, moventis & moti, non satis discriminant causam ab effectu. Similiter, qui nisum aut conatum in motu fingunt, aut idem corpus simul in contrarias partes ferri putant, eâdem idearum confusione, eâdem vocum ambiguitate ludificari videntur.

§51 Juvat multum, sicut in aliis omnibus, ita in scientia de motu accuratam diligentiam adhibere, tam ad aliorum conceptus intelligendos quam ad suos enunciandos: in qua re nisi peccatum esset, vix credo in disputationem trahi potuisse, utrùm corpus indifferens sit ad motum & ad quietem necne. Quoniam enim experientia constat, esse legem naturæ primariam, ut corpus perinde perseveret in *statu motis ac quietis, quamdiu aliunde nihil accidat ad statum istum mutandum.* Et propterea vim interiæ sub diverso respectu esse vel resistentiam, vel impetum, colligitur. Hoc sensu, profecto corpus dici potest sua natura indifferens ad motum vel quietem. Nimirum, tam difficile est quietem in corpus motum, quam motum in quiescens inducere; cum vero corpus pariter conservet statum utrumvis, quid ni dicatur ad utrumvis se habere indifferenter?

§52 Peripatetici pro varietate mutationum, quas res aliqua subire potest, varia motus genera distinguebant. Hodie de motu agentes intelligunt solummodo mótum localem. Motus autem localis intelligi nequit nisi simul intelligatur quid sit *locus*; is vero à neotericis definitur *pars spatii quam corpus occupat,* unde dividitur in relativum & absolutum pro ratione spatii. Distinguunt enim inter spatium absolutum sive verum, ac relativum sive apparens. Volunt scilicet dari spatium undequaque immensum, immobile, insensibile, corpora universa permeans & continens, quod vocant spatium absolutum. Spatium, autem, à corporibus comprehensum, vel definitum, sensibusque adeo subjectum, dicitur spatium relativum, apparens, vulgare.

§53 Fingamus itaque corpora cuncta destrui & in nihilum redigi. Quod reliquum est vocant spatium absolutum, omni relatione quæ à situ & distantiis corporum oriebatur, simul cum ipsis corporibus, sublatà. Porro spatium illud est infinitum, immobile, indivisible, insensibile, sine relatione & sine distinctione. Hoc est, omnia ejus attributa sunt privativa vel negativa: videtur igitur esse merum nihil. Parit solummodo difficultatem aliquam quod extensum sit. Extensio autem est qualitas positiva. Verùm qualis tandem extensio est illa, quæ nec dividi potest, nec mensurari, cujus nullam partem, nec sensu percipere, nec imaginatione depingere possumus? Etenim nihil im imaginationem cadit, quod, ex natura rei, non possibile est ut sensu percipiatur, siquidem imaginatio nihil aliud est quam facultas representatrix rerum sensibilum, vel actu existentium, vel saltem possibilum. Fugit insuper intellectum purum, quum facultas illa versetur tantum circa res spirituales & inextensas, cujusmodi sunt mentes nostræ, earumque habitus, passiones, virtutes & similia. Ex spatio igitur absoluto, auferamus modò vocabula, & nihil remanebit in sensu, imaginatione aut intellectu; nihil aliud ergo iis designatur, quam pura privatio aut negatio, hoc est, merum nihil.

§54 Confitendum omnino est nos circa hanc rem gravissimis præjudiciis teneri, à quibus ut liberemur, omnis animi vis exerenda. Etenim multi, tantum abest quod spatium absolutum pro nihilo ducant ut rem esse ex omnibus (Deo excepto) unicam existiment, quæ annihilari non possit: statuantque illud suapte natura necessariò existere, æternumque esse & increatum, atque adeo attributorum divinorum particeps. Verum enimvero quum certissimum sit, res omnes, quas nominibus designamus, per qualitates aut relationes, vel aliqua saltem ex parte, cognosci, (ineptum enim foret vocabulis uti quibus cogniti nihil, nihil notionis, ideæ vel conceptus subjiceretur.) Inquiramus diligenter, utrum formare liceat ideam ullam spatii illius puri, realis, absoluti, post omnium corporum annihilationem perservantis existere. Ideam porro talem paulo acrius intuens, reperio ideam esse nihili purissimam, si modo idea appellanda sit. Hoc ipse summa adhibita diligentia expertus sum: Hoc alios pari adhibita diligentia experturos reor.

§55 Decipere nos nonnunquam solet, quod aliis omnibus corporibus imaginatione sublatis, nostrum tamen manere supponimus. Quo supposito, motum membrorum ab omni parte liberrimum imaginamur. Motus autem sine spatio concipi non potest. Nihilominus si rem attento animo recolamus, constabit primo concipi spatium relativum partibus nostri corporis definitum: 2°. movendi membra potestatem liberrimam nullo obstaculo retusam: & præter hæc duo nihil. Falso tamen credimus tertium aliquod, spatium, videlicet, immensum realiter existere, quod liberam potestatem nobis faciat movendi corpus nostrum: ad hoc enim requiritur absentia solummodo aliorum corporum. Quam absentiam, sive privationem corporum, nihil esse positivum fateamur necesse est.*

§56 Cæterum hasce res nisi quis libero & acri examine perspexerit, verba & voces parum valent. Meditanti vero, & rationes secum reputanti, ni fallor, manifestum erit, quæcunque de spatio puro & absoluto prædicantur, ea omnia de nihilo prædicare posse. Qua ratione mens humana facillimè liberatur à magnis difficultatibus, simulque ab ea absurditate tribuendi existentiam necessariam ulli rei præterquam soli Deo optimo maximo.

§57 In proclivi esset sententiam nostram argumentis à posteriori (ut loquuntur) ductis confirmare, quæstiones de spatio absoluto proponendo, exempli gratia, utrum sit substantia vel accidens? Utrum creatum vel increatum? & absurditates ex utravis parte consequentes demonstrando. Sed brevitati consulendum. Illud\tamen omitti non debet, quod sententiam hancce Democritus olim calculo suo comprobavit, uti auctor est Aristoteles 1. I. phys. ubi hæc habet; *Democritus solidum & inane ponit principia, quarum aliud quidem ut quod est, aliud ut quod non est esse dicit.* Scrupulum si forte injiciat, quod distinctio illa inter spatium absolutum & relativum à magni nominis philosophis usurpetur, eique quasi fundamento inædificentur multa præclara theoremata, scrupulum istum vanum esse, ex iis, quæ secutura sunt, apparebit.

§58 Ex præmissis patet, non convenire, ut definiamus locum verum corporis, esse partem spatii absoluti quam occupat corpus, motumque verum seu absolutum esse mutationem loci veri & absoluti. Siquidem

---

* Vide quæ contra spatium absolutum disseruntur in libro de principiis cognitionis humanæ, idiomate anglicano, decem abhinc annis edito.

omnis locus est relativus, ut et omnis motus. Veruntamen ut hoc clarius appareat, animadvertendum est, motum nullum intelligi posse sine determinatione aliqua seu directione, quæ quidem intelligi nequit, nisi præter corpus motum, nostrum etiam corpus, aut aliud aliquod, simul intelligatur existere. Nam sursum, deorsum, sinistrorsum, dextrorsum omnesque plagæ & regiones in relatione aliqua fundantur, &, necessario, corpus à moto diversum connotant & supponunt. Adeo ut, si reliquis corporibus in nihilum redactis, globus, exempli gratia, unicus existere supponatur; in illo motus nullus concipi possit; usque adeo necesse est, ut detur aliud corpus, cujus situ motus determinari intelligatur. Hujus sententiæ veritas clarissime elucebit, modo corporum omnium tam nostri quam aliorum præter, globum istum unicum, annihilationem recte supposuerimus.

§59 Concipiantur porro duo globi, & præterea nil corporeum, existere. Concipiantur deinde vires quomodocunque applicari, quicquid tandem per applicationem virium intelligamus, motus circularis duorum globorum circa commune centrum nequit per imaginationem concipi. Supponamus deinde cœlum fixarum creari: subito ex concepto appulsu globorum ad diversas cœli istius partes motus concipietur. Scilicet cum motus natura sua sit relativus, concipi non potuit priusquam darentur corpora correlata. Quemadmodum nec ulla alia relatio sine correlatis concipi potest.

§60 Ad motum circularem quod attinet, putant multi, crescente motu vero circulari, corpus necessario magis semper magisque ab axe niti. Hoc autem ex eo provenit, quod, cum motus circularis spectari possit tanquam in omni momento à duabus directionibus ortum trahens, una secundum radium, altera secundum tangentem; si in hac ultima tantùm directione impetus augeatur, tum à centro recedet corpus

motum, orbita vero desinet esse circularis. Quod si æqualiter augeantur
vires in utraque directione, manebit motus circularis, sed acceleratus
conatu, qui non magis arguet vires recedendi ab axe, quam accedendi
ad eundem, auctas esse. Dicendum igitur, aquam in situla circumac-
tam ascendere ad latera vasis, propterea quod, applicatis novis viribus
in directione tangentis ad quamvis particulam aquæ, eodem instanti non
applicentur novæ vires æquales centripetæ. Ex quo experimento nullo
modo sequitur, motum absolutum circularem per vires recedendi ab axe
motus necessariò dignosci. Porrò, qua ratione intelligendæ sunt voces
istæ, *vires corporum & conatus*, ex præmissis satis superque innotescit.

§61 Quo modo curva considerari potest tanquam constans ex rectis
infinitis, etiamsi revera ex illis non constet, sed quòd ea hypothesis ad
gometriam utilis sit, eodem motus circularis spectari potest, tanquam
à directionibus rectilineis infinitis ortum ducens, quæ suppositio utilis
est in philosophia mechanica. Non tamen ideo affirmandum, impossibile
esse, ut centrum gravitatis corporis cujusvis successive existat in sin-
gulis punctis peripheriæ circularis, nulla ratione habita directionis ullius
rectilineæ, sive in tangente, sive in radio.

§62 Haud omittendum est, motum lapidis in fundâ, aut aquæ in
situlâ circumacta dici non posse motum vere circularem, juxta mentem
eorum qui per partes spatii absoluti definiunt loca vera corporum; cum
sit mirè compositus ex motibus non solum situlæ vel fundæ, sed etiam
telluris diurno circa proprium axem, menstruo circa commune centrum
gravitatis terræ & lunæ, & annuo circa solem. Et propterea, particula
quævis lapidis vel aquæ discribat lineam à circulare longe abhorrentem.
Neque revera est, qui creditur, conatus axifugus, quoniam non respicit
unum aliquem axem ratione spatii absoluti, supposito quod detur tale
spatium: proinde non video quomodo appellari possit conatus unicus,

cui motus vere circularis tanquam proprio & adæquato effectui respondet.

§63 Motus nullus dignosci potest, aut mensurari, nisi per res sensibiles. Cum ergo spatium absolutum nullo modo in sensus incurrat, necesse est ut inutile prorsus sit ad distinctionem motuum. Præterea, determiatio sive directio motui essentialis est, illa vero in relatione consistit. Ergo impossibile est ut motus absolutus concipiatur.

§64 Porrò, quoniam pro diversitate loci relativi, varius sit motus ejusdem corporis, quinimò, uno respectu moveri, altero quiescere dici quidpiam possit: ad determinandum motum verum & quietem veram, quo scilicet tollatur ambiguitas, & consulatur mechanicæ philosophorum, qui systema rerum latius contemplantur, satis fuerit spatium relativum fixarum, cœlo, tanquam quiescente spectato, conclusum adhibere, loco spatii absoluti. Motus autem & quies tali spatio relativo definiti, commodè adhiberi possunt loco absolutorum, qui ab illis nullo symptomate discerni possunt. Etenim imprimantur utcunque vires: sint quicunque conatus: concedamus motum distingui per actiones in corpora exercitas; nunquam tamen inde sequetur, dari spatium illud, & locum absolutum, ejusque mutationem esse motus verus.[5]

§65 Leges motuum, effectusque, & theoremata eorundem proportiones & calculos continentia, pro diversis viarum figuris, accelerationibus itidem & directionibus diversis, mediisque plus minusve resistentibus, hæc omnia constant sine calculatione motus absoluti. Uti vel ex eo patet quod, quum secundum illorum principia qui motum absolutum inducunt, nullo symptomate scire liceat, utrum integra rerum compages quiescat, an moveatur uniformiter in directum, perspicuum sit motum absolutum nullius corporis cognosci posse.

---

[5] reading 'motus verus' here for the nonsensical 'locum verum' which appears in both editions.

§66 Ex dictis patet ad veram motus naturam perspiciendam summopere juvaturum: 1°. Distinguere inter hypotheses mathematicas & naturas rerum. 2°. Cavere ad abstractionibus. 3°. Considerare motum tanquam aliquid sensible, vel saltem imaginibile: mensurisque relativis esse contentos. Quæ si fecerimus, simul clarissima quæque philosophiæ mechanicæ theoremata, quibus reserantur naturæ recessus, mundique systema calculis humanis subjicitur, manebunt intemerata: et motus contemplatio à mille minutiis, subtilitatibus, ideisque abstractis libera evadet. Atque hæc de natura motûs dicta sufficiant.

§67 Restat, ut disseramus de causa communicationis motuum. Esse autem vim impressam in corpus mobile, causam motus in eo plerique existimant. Veruntamen, illos non assignare causam motus cognitam, & à corpore motuque distinctam, ex præmissis constat. Patet insuper vim non esse rem certam & determinatam, ex eo quod viri summi de illa multùm diversa, immo contraria, proferant, salva tamen in consequentiis veritate. Siquidem Newtonus ait vim impressam consistere in actione sola, esseque actionem exercitam in corpus ad statum ejus mutandum, nec post actionem manere. Torricellius cumulum quendam sive aggregatum virium impressarum per percussionem in corpus mobile recipi, ibidemque manere atque impetum constituere contendit. Idem fere Borellus aliique prædicant. At vero, tametsi inter se pugnare videantur Newtonus & Torricellius, nihilominus, quum dum singuli sibi consentanea proferunt, res satis commodè ab utriusque explicatur. Quippe vires omnes corporibus attributæ, tam sunt hypotheses mathematicæ quam vires attractivæ in planetis & sole. Cæterùm entia mathematica in rerum natura stabilem essentiam non habent: pendet autem à notione definientis: unde eadem res diversimodè explicari potest.

§68 Statuamus motum novum in corpore percusso conservari, sive per vim insitam, qua corpus quodlibet perseverat in statu suo, vel quietis, vel motus uniformis in directum:[6] sive per vim impressam, durante percussione in corpus percussum receptam ibidemque permanentem, idem erit quoad rem, differentia existente in nominibus tantùm. Similiter, ubi mobile percutiens perdit, & percussum acquirit motum, parum refert disputare, utrum motus acquisitus sit idem numero cum motu perdito, ducit enim in minutias metaphysicas, & prorsus nominales de identitate. Itaque sive dicamus motum transire à percutiente in percussum, sive in percusso motum de novo generari, destrui autem in percutiente, res eodem recidit. Utrobique intelligitur unum corpus motum perdere, alterum acquirere, & præterea nihil.

§69 Mentem, quæ agitat & continet universam hancce molem corpoream, estque causa vera efficiens motus, eandem esse, propriè & strictè loquendo, causam communicationis ejusdem haud negaverim. In philosophia tamen physicâ, causas & solutiones phænomenôn a principiis mechanicis petere oportet. Physicè igitur res explicatur non assignando ejus causam verè agentem & incorpoream, sed demonstrando ejus connexionem cum principiis mechanicis: cujusmodi est illud, *actionem & reactionem esse semper contrarias & æquales,* à quo, tanquam fonte & principio primario, eruuntur regulæ de motuum communicatione, quæ à neotericis, magno scientarum bono, jam ante repertæ sunt & demonstratæ

§70 Nobis satis fuerit, si innuamus principium illud alio modo declarari potuisse. Nam si vera rerum natura, potius quam abstracta mathesis spectetur, videbitur rectius dici, in attractione vel percussione passionem corporum, quam actionem, esse utrobique æqualem. Exempli

---

[6] Reading 'vel quietis, vel motus uniformis in directum' for 'vel motus, vel quietis uniformis in directum' which appears in both editions.

gratia, lapis fune equo alligatus tantum trahitur versus equum, quantum equus versus lapidem: corpus etiam motum in aliud quiescens impactum, patitur eandem mutationem cum corpore quiescente. Et quoad effectum realem, percutiens est item percussum, percussumque percutiens. Mutatio autem illa est utrobique, tam in corpore equi quam in lapide, tam in moto quam in quiescente, passio mera. Esse autem vim, virtutem, aut actionem corpoream talium effectuum verè & propriè causatricem non constat. Corpus motum in quiescens impingitur, loquimur tamen activè, dicentes illud hoc impellere: nec absurdè in mechanicis, ubi ideæ mathematicæ potius quam veræ rerum naturæ spectantur.

§71 In physica, sensus & experientia, quæ ad effectus apparentes solummodo pertingunt, locum habent; in mechanica, notiones abstractæ mathematicorum admittuntur. In philosophia prima seu metaphysica agitur de rebus incorporeis, de causis, veritate, & existentia rerum. Physicus series sive successiones rerum sensibilium contemplatur, quibus legibus connectuntur, & quo ordine, quid præcedit tanquam causa, quid sequitur tanquam effectus animadvertens. Atque hac ratione dicimus corpus motum esse causam motûs in altero, vel ei motum imprimere, trahere etiam, aut impellere. Quo sensu causæ secundæ corporeæ intelligi debent, nullâ ratione habitâ veræ sedis virium, vel potentiarum actricum, aut causæ realis cui insunt. Porro, dici possunt causæ vel principia mechanica, ultra corpus, figuram, motum, etiam axiomata scientiæ mechanicæ primaria, tanquam causæ consequentium spectata.

§72 Causæ verè activæ meditatione tantum & ratiocinio è tenebris erui quibus involvuntur possunt, & aliquatenus cognosci. Spectat autem ad philosophiam primam, seu metaphysicam, de iis agere. Quod si cuique scientiæ provincia sua tribuatur, limites assignentur, principia

& objecta accuratè distinguantur, quæ ad singulas pertinent, tractare licuerit majore, cum faciltate, tum perspicuitate.

FINIS

DE MOTU;

OR

On the Principle and Nature of Motion, And on the Cause of the Communication of Motions

§1. In the pursuit of truth the most important thing is to beware that poorly understood words do not hinder us: nearly all philosophers warn of this, but few heed the warning. Nevertheless this hardly seems so difficult in matters principally treated by physicists, where sense, experience, and geometrical reasoning have their place. And so setting aside, as much as possible, all prejudice, whether originating in habit of speech or the authority of philosophers, we must diligently examine the very nature of things. Nor indeed should the authority of anyone be valued to the point where his words and terms are prized while nothing clear and certain can be discovered in them.

§2. The contemplation of motion remarkably troubled the minds of the ancient philosophers, from which various exceedingly difficult (not to say absurd) opinions originated, which have now almost fallen into obscurity and so hardly merit that we should dwell upon discussing them in much detail. But among the more recent and sounder philosophers of this age, when they treat of motion, not a few words of the most abstract and obscure signification occur, some of this sort being 'solicitation of gravity', 'conatus', 'dead forces', and so forth.[1] These shroud in darkness writings in other respects most learned, and give rise to opinions no less abhorrent to truth than to common sense. Thus indeed it is necessary, for the sake of truth and not in the interest of refuting others, that these be carefully discussed.

§3. Solicitation and effort or striving apply in truth only to animate things. When they are attributed to other things, they must be taken in a metaphorical sense. But the philosopher should abstain from metaphor. Besides, as anyone who has considered the matter seriously will admit, apart from both affections of the mind and motion of bodies, nothing is clearly and distinctly signified by these words.

§4. As long as heavy bodies are supported by us, we feel in ourselves effort, fatigue, and discomfort. We also percieve in descending heavy bodies an accelerated motion toward the center of the earth: but beyond this we perceive nothing by sense. By reason however we gather that there is some cause or principle of these phenomena, and this is commonly called "gravity." But since the cause of the descent of heavy bodies is unseen and unknown, gravity in this sense cannot properly be called a sensible quality. It is therefore an occult quality. But one can scarcely (and indeed not even scarcely) conceive what an occult quality is, or how any quality could act or effect anything. And so it would be better if, dismissing the occult quality, men attended only to sensibile effects; and if setting aside abstract terms in meditation (however useful they may be in discussion), the mind were fixed on particulars and concrete things, that is on the things themselves.

---

[1] These terms are taken from Leibniz's essay *Specimen Dynamicum* and are part of the Leibnizian theory of forces as explained in Section 1.3 of the "Editor's Introduction."

§5. "Force" is similarly attributed to bodies; but this word is taken as if it signified a known quality, and one as distinct from motion, figure and every other sensibile thing as from every affection of a living thing. Yet anyone examining the matter more closely will agree that this is nothing other than an occult quality. Animal effort and corporeal motion are commonly regarded as the symptoms and measures of this occult quality.

§6. It is therefore obvious that gravity or force are taken in vain as the principle of motion: for how can this principle be more clearly known by being called an occult quality? What itself is occult explains nothing. And let us overlook the fact that this unknown active cause would be better called a substance than a quality. Moreover, 'force,' 'gravity', and words of this sort are more often (and not improperly) taken in concrete, so as to mean the body moved, the difficulty of resisting, &c. Yet when they are used by philosophers to signify certain natures prescinded and abstracted from all these things, which neither are the object of the senses nor can be framed by any force of the intellect or the imagination, then indeed they produce errors and confusion.

§7. But many are led into error because they see that general and abstract terms are useful in discussion and yet do not suficiently understand their meaning. In part these terms were invented by common habit in order to abbreviate speech, and in part they have been devised by philosophers' for instruction: not because they are adapted to the natures things, which are only singular and exist in concrete, but only as they are fit for handing down teachings since they make notions (or at least propositions) universal.

§8. For the most part we suppose that corporeal force is something easy to conceive. But those who have considered the matter more carefuly are of a different opinion, as appears from the remarkable obscurity of the words they use when they try to explain it. Torricelli says that force and impetus are certain abstract and subtle things, and quintessences which are included in corporeal substance as in the magic vase of Circe.* Leibniz likewise, in explaining the nature of force, has this: "Active primitive force, which is the first entelechy corresponds to the soul or substantial form." See the *Acta* of Leipzig.[2] As long as they indulge so far in abstractions, it is necessary that even the greatest men pursue terms endowed with no signification and which are mere shadows of scholastic things. Other examples, and indeed not a few, could be produced from the writings of more recent authors, by which it would be abundantly established that metaphysical abstractions have not everywhere given way to mechanics and experiments, but still make useless trouble for philosophers.

§9. From this source derive various absurdities, one of which is this: "the force of percussion, however small, is infinitely great."[3] Which

---

* Matter is nothing other than an enchanted vase of Circe, which serves as a receptacle of force and the moments of impetus. Force and the impetuses are abstractions so subtle, and quintessences so volatile, that they cannot be enclosed in any other ampules except the innermost substance of natural solids. [This quotation comes from Lecture IV of Torricelli's *Lezioni Accademiche* (Torricelli 1715, 25), in the context of Torricelli's argument that matter itself is completely powerless and inert, serving only as a recepticle of forces.]

[2] Berkeley's reference here is to Leibniz's essay *Specimen Dynamicum*; the sentence he paraphrases reads "Indeed, primitive force (which is nothing but the first entelechy) corresponds to the *soul or substantial form*." (*GM*, 6: 236; *AG*, 119)

[3] It is unclear whether Berkeley is actually quoting from a specific author here. Both Borelli and Torricelli (whom he elsewhere quotes) uphold the doctrine, and Leibniz alludes to it in his discussion of living and dead forces, adding: "And this is what *Galileo* meant when he said, speaking enigmatically, that the force of impact is infinite in comparison with the simple nisus of heaviness." (*GM*, 6: 238; *AG*, 122). But I know of no passage in these authors which corresponds exactly with Berkeley's words here.

surely supposes that gravity is some real quality different from all others, and that gravitation is, as it were, an action of this quality really distinct from motion; but the smallest percussion produces an effect greater than the greatest gravitation without motion. The former surely causes some motion, the latter none. See the experiments of Galileo, and what Torricelli, Borelli, and others have written on the infinite force of percussion.[4]

§10. And yet it must be allowed that no force is immediately felt by itself, nor otherwise known and measured except by its effect; but there is no effect of a dead force or of a simple gravitation in a quiescent body subject to no actual change. There is, however, some effect of percussion. Since, therefore, forces are proportional to effects, we may conclude that there is no dead force. But neither should we conclude on that account that the force of percussion is infinite. For it is not proper to take any positive quantity as infinite on the grounds that it exceeds by an infinite ratio a null quantity or nothing.

§11. The force of gravitation cannot be separated from momentem, but there is no momentum without velocity, since it is mass mulitplied by velocity. Moreover, velocity cannot be understood without motion, and therefore neither can the force of gravitation. Furthermore, no force becomes known except by action, and is measured by the same, but we cannot prescind the action of a body from motion. Therefore, as long as a heavy body changes the shape of lead placed beneath it or the shape of a chord, so long is it moved: yet when it is quiescent it does nothing, or (which is all the same) it is prohibited from acting. In short, these words 'dead force' and 'gravitation', although they are supposed to signify by

---

[4] The works in question here are Galileo (1974, 281-306) Torricelli (1715, Lectures 2-4) and Borelli (1667).

metaphysical abstraction something distinct from what moves, what is moved, from motion and rest, yet in truth this is nothing at all.

§12. If someone were to say that a weight hung from or placed on a chord acts on it, since it prevents it from restoring itself by elastic force, I say that by the same reasoning any lower body acts on an upper body placed upon it, since it prohibits it from descending: and in truth it cannot be said to be an action of a body that it prohibits another body from being in the place which it occupies.

§13. We sometimes feel the pressure of a gravitating body. But that disagreeable sensation arises from the motion of that heavy body communicated to the fibers and nerves of our body and the changing of their position, and to that extent the sensation should be referred to percussion. In these matters we labor under many and weighty prejudices, but they should be subdued, or better thoroughly banished, by keen and repeated meditation.

§14. In order to prove that any quantity is infinite, it should be shown that some finite homogeneous part is contained in it an infinite number of times. But the dead force is not related to the force of percussion as part to whole, but rather as a point to a line, according to those authors who claim the infinite force of percussion.[5] Many things might be added on this matter, but I fear I am being prolix.

§15. By the principles premissed notable disputes which have so much occupied learned men can be solved. An example of these would be that controversy over the proportion of forces.[6] One side, while conceding that momenta, motions, and impetus for a given mass are simply

---

[5] This is another vague reference to Leibniz, Torricelli, and Borelli. All of the authors hold this doctrine, although the use of the term 'dead force' suggests that Berkeley has Leibniz in mind.

[6] The reference here is to the *vis viva* controversy between Leibniz and the Cartesians, as outlined in Section 2.1 of the "Editor's Introduction."

as the velocities, affirms that the forces are as the squares of the velocities. But everyone sees that this opinion supposes the force of a body to
be distinguished from momentum, motion, and impetus, and it collapses
when this supposition is removed.

§16. To make it appear still more clearly that a certain remarkable
confusion has been introduced into the doctrine of motion by metaphysical abstractions, let us see how much famous men's thoughts on force
and velocity differ. Leibniz confuses impetus with motion.[7] According
to Newton impetus is in truth the same as the force of inertia.[8] Borelli
asserts that impetus is nothing other than a degree of velocity.[9] Some
would have impetus and conatus differ from one another, others hold
that they do not. Many understand the motive force to be proportional
to motion, but some hold that there is some force beyond the motive

---

[7] Leibniz identified impetus with the infinite sum of elementary nisus induced by motion: ". . . *nisus* is twofold, that is, elementary or infinitely small,
which I also call *solicitation*, and that which is formed by the continuation or
repetition of elementary nisus, that is, *impetus* itself." (*GM*, 6: 238; *AG*, 121)
Elsewhere he asserts "*impetus* is the product of bulk [moles] of a body and
its velocity, whose quantity is what the *Cartesians* usually call quantity of motion." (*GM*, 6: 237; *AG*, 120) It is unclear why Berkeley reads this as confusing
impetus with motion.

[8] In commenting on Definition III of the *Principia*, Newton writes: "A body,
from the inert nature of matter, is not without difficulty put out of its state of
rest or motion. Upon which account, this *vis insita* may, by a most significant
name, be called inertia (*vis inertiæ*) or force of inactivity. But a body only
exerts this force when another force, impressed upon it, endeavors to change its
condition; and the exercise of this force may be considered as both resistance
and impulse; it is resistance so far as the body, for maintaining its present state,
opposes the force impressed; it is impulse so far as the body, by not easily giving
way to the impressed force of another, endeavors to change the state of that
other. Resistance is usually ascribed to bodies at rest, and impulse to those
in motion; but motion and rest, as commonly conceived, are only relatively
distinguished; nor are those bodies always truly at rest, which commonly are
taken to be so." Newton (*Principia*, 1: 2)

[9] "Igitur, in ipso motu prout ejus intensio consideratur, concipi debet vis illa
& energia celeritatis, qua corpus movetur, quae in summa nil aliud est quam
mensura gradus velocitatis ejus, atque hujusmodi vis nuncupari solet impetus."
(Borelli 1667, 3)

force, and that it is to be measured differently, in as much as they think it varies as the square of the velocity multiplied by the mass.[10]

§17. 'Force', 'gravity', 'attraction', and words of this sort are useful for reasonings and computations concerning motion and bodies in motion, but not for understanding the simple nature of motion itself, or for designating so many distinct qualities. As for attraction, it is clear that this was employed by Newton, not as a true and physical quality, but only as a mathematical hypothesis.[11] And indeed Leibniz, in distinguishing elementary nisus or solicitation from impetus, confesses that these things are not to be found in things themselves in nature, but must be made by abstraction.[12]

§18. The account of the composition and resolution of any direct forces into oblique ones by the diagonal and sides of a parallelogram is a similar case.[13] These things serve mechanics and computation: but it is one thing to serve computation and mathematical demonstrations, and another to exhibit the nature of things.

§19. Among the more recent thinkers, many are of the opinion which holds that motion is neither destroyed nor generated anew, but rather the same quantity of motion always remains. Indeed Aristotle once posed this question, whether motion is made and passes away, or

---

[10] Another reference to the *vis viva* controversy.

[11] Presumably, Berkeley means to recall Newton's comments on Definition VIII of the *Principia*: "I here design only to give a mathematical notion of those forces, without considering their physical causes and seats." (*Principia*, 1: 5)

[12] Leibniz remarks: "Nevertheless, I wouldn't want to claim on these grounds that these mathematical entities are really found in nature, but I only wish to advance them for making careful calculations through mental abstraction." (*GM*, 6: 238; *AG*, 121)

[13] Berkeley's reference here is to Corollary I, Book I of the *Principia*: "*A body, acted on by two forces simultaneously, will describe the diagonal of a parallelogram in the same time as it would describe the sides by those forces separately.*" (*Principia*, 1: 14)

whether it exists from eternity. *Physics*, Book 8.[14] That sensible motion perishes is plain to the senses, but it seems that these thinkers will have it that the same impetus and nisus, or the same sum of forces remains. Whence Borelli affirms that the force in percussion is not diminished but expanded, and even that contrary impetus are received and retained in the same body.[15] Likewise, Leibniz contends that nisus is everywhere and always in matter and, where it is not evident to the senses, it is understood by reason.[16] But it must be admitted that these things are abstract and obscure, and of nearly the same sort as substantial forms and entelechies.

§20. All those who in explaining the cause and origin of motion make use of a Hylarchic principle, or the need of nature, or its appetite, or lastly of a natural instinct, are to be judged as having said something rather than to have thought anything.[17] Nor is there much difference

---

[14] "Was there ever a becoming of motion before which it had no being, and is it perishing again so as to leave nothing in motion? Or are we to say that it never had any becoming and is not perishing, but always was and always will be? Is it in fact an immortal never-failing property of things that are, a sort of life as it were to all naturally constituted things?" (250^b 10-14, 1: 418)

[15] "In percussione vis motiva impellitis non minuitur, neque de novo ulla in projecto producitur, sed tantummodo expanditur, ita ut una eis pars in percutiente remaneat, reliqua vero in corpus percussum communicietur." Borelli (1667, 48)

[16] "This *nisus* frequently presents itself to the senses and, in my judgement, is understood by reason to be everywhere in matter, even where it is not obvious to sense." (*GM*, 6: 235; *AG*, 118)

[17] This is a reference to the theories of the Cambridge Platonists Henry More and Ralph Cudworth. Cudworth, in Book I of his *True Intellectual System* asserts: "But since it appears plainly, that *Matter* or *Body* cannot *Move it self*; either the *Motion* of all Bodies, mush have no manner of *Cause*, or else must there of necesity be some other *Substance* besides *Body*, such as is *Self-active* and *Hylarchical*, or hath a *Natural Power*, of *Ruling over Matter*." (Cudworth [1678] 1978, 668) It is probable, however, that Berkeley has picked up the reference from Leibniz, who claims "However, even though I admit an active and, so to speak, vital principle superior to material notions everywhere in bodies, I do not agree with *Henry More* and other gentlemen distinguished in piety and ability, who use an Archaeus (unintelligible to me) or hylarchic principle even for dealing with the phenomena, as if not everything in nature

between them and those who have supposed that "the parts of earth are self-moving, and there are even spirits implanted in them corresponding to forms,"* in order to assign a cause for the acceleration of descending heavy bodies; or he who said that "in body beyond solid extension there is need to posit something else from which the consideration of forces might arise."† For indeed all of these either say nothing particular and determinate, or if it is something, it would be as difficult to explain as that very thing which it was adduced to explain.

§21. In illuminating nature it is vain to adduce things which are neither evident to the senses nor intelligible to reason. Let us therefore see what sense, what experience, and lastly what reason resting upon them recommend. There are two supreme classes of things, body and mind. We know by sense something extended, solid, mobile, figured, and endowed with other qualities which meet the senses; but we know by a certain internal consciousness the sentient, percipient, intelligent thing. Further, we discern that these things plainly differ from one another and are widely heterogeneous. I speak of things known, for it is useless to speak of things unknown.

§22. All that which we know and have given the name 'body' contains nothing in itself which could be the principle or efficient cause of

---

can be explained mechanically," (*GM*, 6: 242; *AG*, 125-6)

* Borelli. [This passage comes from Prop. 87 of Borelli's *De Vi Percussionis*. Borelli rejects the opinion that acceleration is caused by the approach of a heavy body toward the Earth, and concludes that the parts of the Earth are self-moving. See Borelli (1667, 180-1)]

† Leibniz. [This is yet another reference to Leibniz's *Specimen Dynamicum*. Leibniz asserts: "one can establish that something should be posited in body over and above size and impenetrability, something from which the consideration of forces arises." (*GM* 6: 241; *AG*, 124) Berkeley misquotes Leibniz, reading 'solid extension' for 'size and impenetrability,' but the reference is otherwise clear.]

motion; for indeed impenetrability, extension, and figure include or connote no power of producing motion. On the contrary, reviewing singly not only these but any other qualities of body, whatever they might be, we will see that they are all in fact passive and there is nothing active in them which could in any way be understood as the source and principle of motion. As for gravity, we have already shown above that by that term is signified nothing known or distinct from this sensible effect whose cause is sought. And surely when we when we call a body heavy, we understand nothing else except that it is borne downward, and we do not think at all of the cause of this sensible effect.

§23. And so regarding body we may boldly declare as established fact that it is not the principle of motion. Then if anyone should contend that the word 'body' also contains in its meaning an occult quality, virtue, form, or essence beyond solid extension and its modifications, leave him to the useless business of disputing without ideas and abusing words expressing nothing distinctly. But the sounder way of philosophizing seems to be to abstain as far as possible from abstract and general notions (if such things which cannot be understood can be called notions).

§24. We know whatever is contained in the idea of body: but it is agreed that what we know in body is not a principle of motion. Those who further imagine something unknown in body, of which they have no idea, and which they call the principle of motion are in fact saying nothing more than that the principle of motion is unknown. But it is annoying to dwell any longer on such subtlties.

§25. Besides corporeal things there is another class, that of thinking things. And that in these there is a power of moving bodies we have learned from our own experience, since our minds may at will excite and stop the motion of our limbs, whatever the ultimate explanation of this may be. This much is certainly agreed, that bodies are moved at the will of the mind, and it can thus quite appropriately be called a principle of motion; a particular and subordinate one indeed, and one which itself depends on the first and universal principle.

§26. Heavy bodies are borne downward, although acted upon by no apparent impulse, but we ought not to judge on that account that the principle of motion is contained in them. Aristotle ascribes this reason in the matter: "heavy and light things," he says "are not moved by themselves, for that would be characteristic of life, and they could stop themselves."[18] All heavy things seek the center of the earth by one and the same certain and constant law, nor is there observed in them a principle or any faculty of stopping this motion, of decreasing it, of augmenting it except in a fixed proportion, or lastly of changing it in any way: they behave only passively. Again, the same should be said of percussive bodies, speaking strictly and accurately. These bodies as long as they are being moved, and also in the moment of percussion, behave passively, and just as when they are at rest. An inert body acts just as a moving body, if the matter is expressed truthfully: and this is what Newton recognizes when he says that the force of inertia is the same as impetus.[19] But an inert body and one at rest do nothing, therefore neither does a moving body.

---

[18] "When these things are in motion to positions the reverse of those they would properly ocupy, their motion is violent: when they are in motion to their proper positions – the light thing up and the heavy thing down – their motion is natural; but in this case it is no longer evident, as it is when the motion is unnatural, whence their motion is derived. It is impossible to say that their motion is derived from themselves: this is a characteristic of life and peculiar to living things. Further, if it were, it would have been in their power to stop themselves. . . . " (255^a 3-7; 1: 426)

[19] This is another reference to Definition III of the *Principia*. See Newton (*Principia*, 1: 2)

§27. In fact, a body persists in either state, whether motion or rest. But this persistence is no more to be called an action of the body, than its existence is called its action. Persistence is nothing more than the continuation in the same mode of existing, which cannot properly be called an action. But the resistence we experience in stopping a moving body we imagine to be its action, misled by meer appearance. For in truth this resistence which we sense is a passivity [*passio*] in us, nor does it prove that the body acts, but that we suffer: it is certainly agreed that we would have suffered the same, whether the body moves itself or is impelled by another principle.

§28. Action and reaction are said to be in bodies; and this is quite convenient in mechanical demonstrations. But we should beware not to suppose for this reason that some real power is in them which is the cause or principle of motion. For indeed these words are to be understood in the same way as the word 'attraction'; and just as this is only a mathematical hypothesis and not a physical quality, the same should be understood of these words, and for the same reason. For just as in the mechanical philosophy the truth and use of the theorems of the mutual attraction of bodies remain firm, since they are solely founded on the motion of bodies, whether this motion is supposed to be caused by the action of bodies mutually attracting each other or by the action of some agent distinct from bodies impelling and controlling them; by a similar reasoning, whatever is said of the rules and laws of motion, and also of the theorems deduced from them, remains unshaken, so long as sensible effects and reasoning based on them are granted, whether we suppose this action itself or the force causing these effects to be in body or in an incorporeal agent.

§29. Let extension, solidity, and figure be taken away from the idea of body and nothing will remain. But these qualities are indifferent to motion, nor do they have anything in them which could be called the principle of motion. This is clear from our ideas themselves. If, therefore, the word 'body' signifies that which we conceive, it is plainly agreed that the principle of motion cannot be sought in body, for surely no part or attribute of it is a true efficient cause which produces motion. But to introduce a word and conceive nothing is unworthy of a philosopher.

§30. There is a thinking, active thing which we experience as the principle of motion in ourselves. This we call "soul," "mind," or "spirit." There is also an extended, inert, impenetrable, mobile thing, which is totally distinct from the former and constitutes a new genus. The first of all to grasp how great a difference there is between thinking and extended things was Anaxagoras, a man of great wisdom, when he declared that the mind has nothing in common with bodies, as is established in the first book of Aristotle's *De Anima*.[20] Among more recent thinkers, Descartes has best observed the same point. What is made clear by him, others have rendered awkward and difficult by obscure terms.

§31. From what has been said it is manifest that those who affirm that active force, action, and the principle of motion are truly in bodies embrace an opinion based on no experience, and they add to it with obscure and very general terms, nor do they adequately understand what they themselves mean. On the other hand, those who hold that mind is the principle of motion advance an opinion secured by personal

---

[20] The reference here is somewhat unclear. Presumably, Berkeley is recalling Aristotle's report that "[A]ll those who admit one cause or element, make the soul also one (e.g. fire or air), while those who admit a multiplicity of principles make the soul also multiple. The exception is Anaxagoras; he alone says that thought is impassible and has nothing in common with anything else." (405^b 15-20; 1: 646)

experience and fully approved in the judgements of the most learned men of all ages.

§32. Anaxagoras was the first who introduced mind (τὸν νουν) which impressed motion on inert matter, an opinion which Aristotle also approves and confirms in many ways, frankly pronouncing that the first mover is immoveable, indivisible, and has no magnitude. For to say that every mover must be moveable, he rightly observes, is the same as if one were to say that every builder must be capable of being built. See *Physics*, Book 8.[21] Plato, moreover, in the *Timæus* teaches that this corporeal machine, or the visible world, is put in motion and animated by a mind, which eludes all sense.[22] And indeed even today the Cartesian philosophers acknowledge that the principle of natural motions is God.[23] And Newton everywhere clearly intimates that not only is motion initially brought about by the will of God, but that the the mundane system is still moved by this same action. This is consistent with the sacred scriptures: this is fully approved by the opinion of the scholastics. For although the peripatetics teach that nature is the principle of motion

---

[21] In dismissing the possibility that everything could be in motion, Aristotle contends that "Still more unreasonable is the consequence that, since everything that is moved is moved by something that is itself moved, everything that has a capacity for causing motion is capable of being moved: i.e. it will have a capacity for being moved in the sense in which one might say that everything that has a capacity for making healthy has a capacity for being made healthy, and that which has a capacity for building has a capacity for being built, either immediately or through one or more links." (257^a 14-18; 1: 429)

[22] There is no specific passage in the *Timæus* which expresses exactly this thought, although it is clearly implied in the dialogue as a whole.

[23] As, for example, when Descartes declares in his conservation principle that God is the ultimate cause of all motion and preserves the same quantity of motion in the universe.

and rest, yet they interpret *natura naturans* as God.[24] They undoubtedly understand that all the bodies of this mundane system are moved by an all-powerful mind, according to a certain and constant ratio.

§33. But those who attribute a vital principle to bodies, imagine something obscure and ill-suited to things. For what else is it to be endowed with a vital principle, except to live? and what is it to live, except to move, stop, and change position? But the most learned philosophers of the present age take it as an indubitable principle, that every body remains in its state, either of rest or uniform motion in a right line, except in so far as it is compelled from without to change this state.[25] In contrast, we sense that in the mind there is the faculty of changing both its own state and that of other things; the mind is on this account properly called vital and completely distinguished from bodies.

§34. More recent thinkers consider motion and rest as two states of existence in bodies, in either of which every inert body remains by its nature when there is no external force acting on it. From this, one may gather that the cause of motion and rest is the same as that of the existence of bodies. Nor indeed does it seem that another cause of the successive existence of bodies in diverse parts of space is to be sought than that from which derives the successive existence of the same body in different parts of time. But to treat of the good an great God, ceator and preserver of all things, and to demonstrate how all things depend on the supreme and true Being, although it is the most excellent part of human knowledge, seems rather more to belong to first philosophy or

---

[24] The scholastic doctrines of nature distinguished between *natura naturans* and *natura naturata*; the former is, roughly, nature as the creating force while the latter is nature as that which is created. In the first sense, God can be identified with nature.

[25] The "learned philosophers of the present age" are clearly those who accept the principle of inertia.

methaphysics and theology than to natural philosophy, which today is almost completely confined to experiments and mechanics. And so natural philosophy either supposes knowledge of God, or borrows it from some superior science. Although it is most true that the investigation of nature supplies the higher sciences with excellent arguments for illustrating and proving the wisdom, goodness, and power of God.

§35. Because these things are not suffiently understood, some unjustly repudiate mathematical principles of physics, evidently on the pretext that they do not assign the true efficient causes of things. When in fact it is the concern of the physicist or mechanician to consider only the rules, not the efficient causes, of impulse or attraction, and, in a word, to set out the laws of motion: and from the established laws to assign the solution of a particular phenomenon, but not an efficent cause.

§36. It will be of great importance to have considered what a principle properly is and in what sense this word is to be understood by philosophers. Now the true, efficient, and conserving cause of all things is most rightly called their source and principle. But the principles of experimental philosophy are properly called the foundations upon which rests, or the sources from which derives (I say not the existence, but) our knowledge of corporeal things, and these foundations are sense and experience. Similarly, in the mechanical philosophy, those things are to be called principles in which the whole discipline is founded and contained: those primary laws of motion which are proved by experiments, refined by reason, and rendered universal. These laws of motion are appropriately called principles, since from them dreive both general mechancial theorems as well as particular explanations of the phenomena.

§37. Then certainly something can be said to be explained mechanically when it is reduced to these most simple and universal principles, and its harmony and connection with them is shown by accurate reasoning. For once the laws of nature are found, then the philosopher is to show that from the constant observance of these laws, that is from these principles, any phenomena necessarily follow. This is what it is to explain and solve the phenomena and to assign their cause, that is the reason why they occur.

§38. The human mind delights in extending and expanding its knowledge. But for this end general notions and propositions must be formed, in which particular propositons and knowledge are in some way contained, which are then (and only then) believed to be undersood. This is well known to geometers. In mechanics also notions are premissed, that is definitions and first and general propositions about motion, from which more remote and less general conclusions are later deduced by the mathematical method. And as by the application of geometrical theorems the particular magnitudes of bodies are measured, so also by the application of the universal theorems of mechanics the motions of any parts of the mundane system, and the phenomena which depend upon these motions, become known and are determined: and this is the only goal at which the physicist should aim.

§39. And just as geometers for the sake of their discipline contrive many things which they themselves can neither describe, nor find in the nature of things, for just the same reason the mechanician employs certain abstract and general words, and imagines in bodies force, action, attraction, solicitation, &c. which are exceedingly useful in theories and propositions, as also in computations concerning motion, even if in the very truth of things and in bodies actually existing they are sought in vain, no less than those things geometers frame by abstraction.

§40. In truth, we perceive nothing by sense except effects or sensible qualities, and corporeal things are completely passive, whether they be in motion or at rest: and reason and experience advise us that there is nothing active except for the mind or soul. Whatever is imagined beyond these should be judged to be of the same sort as other hypotheses and mathematical abstractions; this should be impressed deeply in the mind. If this is not done, we can easily lapse back into the obscure subtlety of the scholastics, which corrupted philosophy for so many ages like some dire plague.

§41. The mechanical principles and universal laws of motions, or of nature, happily discovered in the last century and treated of and applied with the aid of geometry, have cast a remarkable light on philosophy. But metaphysical principles and the real efficient causes of the motion and existence of bodies or of corporeal attrributes in no way pertain to mechanics and experiments, nor can they shed any light on them, except in so far as by being known beforehand they may serve to set the limits of physics, and so to remove foreign difficulties and questions.

§42. Those who derive the principle of motion from spirits understand by the word 'spirit' either a corporeal or incorporeal thing: if they mean a corporeal thing, however subtle, yet the difficulty recurs: if they mean an incorporeal thing, however true this may be, yet it does not properly pertain to physics. For if someone were to extend natural philosophy beyond the limits of experiments and mechanics, so that it would embrace knowledge of incorporeal and unextended things, the broader interpretation of the term admits treating of the soul, mind or vital principle. But it will be more fitting, following widely received practice, so to distinguish among the sciences, that each is circumscribed within its proper bounds, and the natural philosopher will be wholly concerned with experiments, the laws of motion, mechanical principles, and reasonings deduced from these. But whatever he may advance on other matters, let him refer any such claim to a higher science. For from the knowledge of the laws of nature follow the most beautiful theories, as well as mechanical devices useful in life. But from the knowledge of the Author of Nature Himself arise considerations of the very highest order, but these are metaphysical, theological, and moral.

§43. So far concerning the principles: now we must speak speak of the nature of motion, and indeed this, since it is clearly perceived by the senses, is not rendered obscure so much by its own nature as by the learned comments of philosophers. Motion never meets the senses without corporeal mass, space, and time. However, there are some who desire to contemplate motion as a certain simple and abstract idea, and separated from every other thing. But this most tenuous and subtle idea eludes the acuteness of the intellect, as any one can test for himself by meditation. Hence arise the great difficulties over the nature of motion, and definitions far more obscure that this thing which they should illustrate. Of this kind are the definitions of Aristotle and the scholastics, who say that motion is the action "of a mobile thing, so far as it is mobile, or the act of being in potentia so far as it is in potentia."[26] And also of this kind is the definition of a famous man in recent times, who asserts that "there is nothing real in motion beyond a momentary some-

---

[26] Aristotle's definition motion first appears in Book III of the *Physics*, "It is the fulfilment of what is potential when it is already fulfilled and operates not as itself but as movable, that is motion." (201ᵃ 28-30; 1: 343) This is repeated in various forms throughout the *Physics* and scholastic commentaries upon it.

thing which must consist in a force striving toward change."[27] Again, it is agreed that the authors of these and similar definitions had it in mind to explicate the abstract nature of motion, apart from any consideration of time and space, but I do not see how this abstract quintessence of motion (as I may call it) can be understood.

§44. Not content with this, they go further and divide and distinguish from one another the parts of motion itself, the distinct ideas of which they try to form, as if of entities truly distinct. For there are those who distinguish movement [motio] from motion [motus], regarding the former as an instantaneous element of motion.[28] Moreover, they would have velocity, conatus, force, and impetus to be so many things differing in essence, each of which is presented to the intellect through its own abstract idea, separated from all the others. But there is no need to remain in discussions of these things, if what we set forth above is accepted.

§45. Many also define motion by *transition*, forgetting of course that transition itself cannot be understood without motion, and should be defined by it.[29] So very true it is that definitions, as they shed light on some things, in turn shroud others in darkness. And indeed, anything we perceive by sense could hardly have been made clearer or better known

---

[27] Another Leibniz citation from the *Specimen Dynamicum*: "For, strictly speaking, motion (and likewise time) never really exists, since the whole never exists, inasmuch as it lacks coexistent parts. And furthermore, there is nothing real in motion but a momentary something which must consist in a force striving toward change." (*GM*, 6: 235; *AG*, 119)

[28] The doctrine in question is Leibniz's. See (*GM*, 6: 237; *AG*, 120) and Section 1.3 of the "Editor's Introduction" for more on the relevant distinction.

[29] The reference here may be to the first Chapter of Borelli's *De Vi Percussuionis*, which declares: "Erit igitur motus localis transitus succesivus ab uno ad alium locum in aliquo determinatio tempore excurrendo succesivis contactibus partes omnes loci, seu spatii transacti sese consequentes." (Borelli 1667, 1-2).

by any definition. Enticed by the vain hope of doing this, philosophers have rendered easy things difficult, and entangled their minds in difficulties, which for the most part they themselves have engendered. From this fondness for defining, as well as abstracting, many very subtle questions concerning motion and other things have arisen, and as these are of no use, they have tortured the minds of men in vain, so that Aristotle even frequently confesses that motion is "a certain act difficult to know,"[30] and some of the ancients became so practiced in these trifles that they denied the existence of motion altogether.[31]

§46. But it is annoying to dwell on such minutiae. It is enough to have indicated the sources of the solutions, to which indeed this may be added: that those things concerning the infinite division of time and space which are taught in mathematics have, by the very nature of the case, introduced paradoxes and thorny theories (as are all those in which infinity is concerned) into speculations about motion. But anything of this kind is something which motion has in common with space and time, or rather is something which it has taken over from them.

§47. And just as on the one hand too much abstraction or divison of things truly inseparable renders the nature of motion preplexed, so on the other hand does composition (or rather confusion) of things very diverse. For it is usual to confound motion with the efficient cause of motion. Whence it happens that motion is treated as if it were somehow

---

[30] In Book III of the *Physics*, Aristotle declares "The reason why motion is thought to be indefinite is that it cannot be classed as a potentiality or as an actuality – a thing that is merely *capable* of having a certain size is not necessarily undergoing change, nor yet a thing that is *actually* of a certain size, and motion is thought to be a sort of *actuality*, but incomplete, the reason for this view being that the potential whose actuality it is is incomplete. This is why it is hard to grasp what motion is." (201^b 26-33; 1: 344)

[31] Presumably a reference to the paradoxes of Zeno, as well as the Parmenidean doctrine that change is impossible.

twofold, having one aspect open to the senses, the other shrouded in the darkest night. Whence obscurity, confusion, and various paradoxes concerning motion draw their origin, when that is falsely attributed to the effect which in truth can belong only to the cause.

§48. Hence arises the opinion that the same quantity of motion is always conserved; which is false, as anyone will easily agree, unless it is understood of force and the power of the cause, or that cause is called nature or νοῦς, or some such agent. Aristotle indeed, in Book 8 of the *Physics* seems to have understood a vital principle rather than the external effect, or change of place, when he asks "whether motion is created and corrupted, or whether it is truly present in all things from eternity like immortal life?"[32]

§49. Hence also it is that many suspect that motion is not a mere passivity [*passio*] in bodies. But if we understand it as that which in the motion of bodies is presented to the senses, no one can doubt that this is completeley passive. For what is there in the successive existence of a body in different places, which could involve action or be something other than a bare and inert effect?

§50. The Peripatetics, who say that motion is one act of two things, the mover and the moved, do not sufficiently discriminate the cause from the effect. Similarly, those who imagine nisus or conatus in motion, or that the same body is simultaneously borne in two contrary directions, appear to be deluded by the same confusion of ideas and the same ambiguity of terms.

---

[32] "Was there ever a becoming of motion before which it had no being, and is it perishing again so as to leave nothing in motion? Or are we to say that it never had any becoming and is not perishing, but always will be? Is it in fact an immortal never-failing property of things that are, a sort of life as it were to all naturally constituted things?" (250^b 10-14; 1: 418)

§51. As in all other things, so also in the science of motion it is of
great use to exercise careful attentiveness, as much to the understanding
of the concepts of others as to the ennunciation of one's own: unless there
has been a failing in this respect, I scarcely believe that the dispute could
have dragged on over whether or not a body is indifferent to motion and
rest. For since it is established by experience that it is a primary law
of nature that a body persists "in a state of motion or rest, as long as
nothing happens from elsewhere to change that state,"[33] and therefore
it is inferred that the force of inertia is under different aspects either
resistence or impetus, in this sense a body can indeed be said to be
by nature indifferent to motion and rest. Certainly it is as difficult to
impart rest to a moving body as it is to impart motion to a body at
rest; but when a body equally preserves either state, why should it not
be said to be indifferently disposed to both?

§52. The Peripatetics used to distinguish a variety of motions by
the variety of changes which any thing could undergo. Today those
who are concerned with motion understand only local motion. But they
deny that local motion can be understood unless in is also understood
what *location* [*locus*] is. This is indeed defined by the moderns as "a
part of space which a body occupies,"[34] whence location is divided into
absolute and relative according to which space is understood. For they
distinguish between absolute or true space, and relative or apparent
space.[35] Indeed they maintain that there is a space on all sides, immense,
immobile, insensible, permeating and containing every body, which they

---

[33] Presumably, this is a Berkeleyan paraphrase of the law of inertia, although
there is no source which uses exactly these words.

[34] For example, in Newton's Scholium to the Definitions in the *Principia*, we
read "Place is a part of space which a body takes up, and is according to the
space, either absolute or relative." (*Principia*, 1: 6)

[35] They, of course, are Newton and the Newtonians.

call absolute space. But space comprehended or defined by bodies and so subjected to the senses is called relative, apparent, and vulgar.

§53. And so let us imagine that all bodies have been destroyed and reduced to nothing. What remains they call absolute space, all relation which arose from the position and distances of bodies having been removed along with the bodies themselves. Now this space is infinite, immobile, indivisible, insensible, without relation and without distinction. That is, all of its attributes are privative and negative: it seems therefore to be merely nothing. The only difficulty arises from the fact that it is something which is extended, and extension is a positive quality. But what kind of extension is that which can neither be divided, nor measured, which has no parts, which we can neither perceive by sense nor depict in the imagination? For nothing enters the imagination which, from the nature of the thing, is impossible to be perceived by sense, since indeed the imagination is nothing other than a representative faculty of sensible things, either actually existing, or at least possible. It also evades the pure intellect, which faculty is concerned only with spiritual and unextended things, such as our minds, their states, passions, powers, and such like.[36] Therefore let us take from absolute space just the words, and nothing will remain in sense, imagination, or intellect; therefore they designate nothing, except pure privation or negation, that is, merely nothing.

§54. It must be frankly confessed that we are in the grip of the most serious prejudices in this matter, and in order to free ourselves from them every effort of the mind must be exerted. And indeed many are

---

[36] The faculty of "pure intellection" was postulated by many philosophers of the period, who treated it as a mental faculty capable of contemplating nonsensory ideas. Berkeley is suspicious of any such theory of pure intellection, and suggests that it could, at best, deal with ideas of "spiritual things." This view is also set forth in a letter from Berkeley to Jean LeClerc; see *Works*, 8: 49-50.

so far from considering absolute space as nothing that they are led to
regard it as the only thing (except God) which cannot be annihilated:
and they hold that it exists necessarily of its own nature, and that it
is eternal, uncreated, and even shares the divine attributes.[37] And yet
surely it is most certain that all things which we designate by names are
known at least in part by qualities or relations (since it would be foolish
to use words to which nothing known, no notions, ideas, or concepts
were attached). Let us then carefully inquire whether any idea can be
formed of this pure, real, absolute space which continues to exist after
the annihilation of all bodies. And yet when I inspect such an idea
somewhat more closely, I find it to be the idea of the purest nothing, if
it can even be called an idea. This I have experienced after paying the
most careful attention: I expect that others will experience it if they
give the matter similar attention.

§55. We may sometimes be deceived by the fact that when in imagination we suppose all other bodies to be removed, we still suppose our
own body to remain. Under this supposition, we imagine the freest
motion of our limbs on all sides. But motion without space cannot be
conceived. Nevertheless if we consider the matter more attentively, it
will be clear that we conceive first a relative space marked out by the
parts of our body, secondly, the fully free power of moving our limbs
impeded by no obstacle, and nothing beyond this. And yet we falsely
believe that there is some third thing, namely immense space really existing, which invests us with the free power of of moving our body: but

---

[37] Berkeley presumably has in mind Joseph Raphson's essay *Spatio Reali*,
Chapters 5 and 6 of which set out various divine attributes shared by absolute
space. See Raphson (1697).

for this is required only the absence of other bodies. And we must admit that this absence, or privation of bodies, is nothing positive.*

§56. But unless one has examined these things freely and carefully, the words and terms are of little worth. Yet to one who meditates and reflects, if I am not mistaken, it will be manifest that whatever is predicated of pure and absolute space can as well be predicated of nothing. By this reasoning the human mind is most easily freed from great difficulties and from the absurdity of attributing necessary existence to any thing beyond the good and great God alone.

§57. It would be easy to confirm our opinion by arguments drawn (as they say) *a posteriori*, by proposing questions concerning absolute space, for example, whether it is a substance or accident? whether it is created or uncreated? and then showing the absurdities following from either answer. But I must be brief. It ought not to be omitted, however, that Democritus formerly affirmed this account in his opinion, as the authority of Aristotle attests in Book I of the *Physics*, where he writes "Democritus takes as principles the solid and the void, of which he says that the one is as what is, the other as what is not."[38] But if by chance someone should raise the doubt that this distinction between absolute and relative space is used by philosophers of great name, and that many famous theorems have been based on it as a foundation, it will appear from what follows that this doubt is in vain.

§58. From what has been said it is clear that it is not fitting that we should define the true location of a body to be the part of absolute

---

* See what is set forth against absolute space in the book on *The Principles of Human Knowledge*, in the English tounge, published ten years ago. [In particular, §§112-116 contain a polemic against absolute space which has much in common with that presented here.]

[38] "The same is true of Democritus also, with his plenum and void, both of which exist, he says, the one as being, the other as not being." (188^a 22-23; 1: 321)

space which the body occupies, and true or absolute motion to be a change in true or absolute location. Since all location is relative, so also is all motion. But that this may appear yet more clearly, it should be noticed that no motion can be understood without some determination or direction, which itself cannot be understood unless besides the moving body, our own body or some other is also understood to exist. For up, down, left, right, and all places and regions are founded on some relation, and necessarily connote and suppose a body distinct from the body moved. And so when all other bodies are supposed to be reduced to nothing, and a solitary globe, for example, to exist; no motion can be conceived in it; so necessary it is that some other body be given, the position of which is understood to determine the motion. The truth of this opinion will shine forth most clearly, when we rightly suppose the annihilation of all bodies, both our own as well as all others besides this solitary globe.

§59. Then let two globes be conceived to exist, and beyond them nothing corporeal.[39] Let forces then be conceived to be applied in some way. Whatever we ultimately understand by the application of forces, a circular motion of the two globes about a common center cannot be conceived by the imagination. Then let us suppose the heaven of fixed stars to be created: suddenly from the conception of the approach of the globes to different parts of that heaven, motion will be conceived. For of course as the nature of motion itself is relative, it could not have

---

[39] Here, Berkeley is responding to Newton's arguments for absolute space. Newton had insisted that "if two globes, kept at a given distance one from the other by means of a cord that connects them, were revolved about their common centre of gravity, we might, from the tension of the cord, discover the endeavor of the globes to recede from the axis of their motion, and from thence we might compute the quantity of their circular motion." (*Principia*, 1: 12)

been conceived before correlated bodies were given. Nor can any other relation be conceived without correlatives.

§60. As regards circular motion, many think that as true circular motion increases, the body necessarily endeavours to recede more and more from the axis. This belief arises from the fact that since circular motion can be seen as taking its origin at every moment from two directions, one along the radius and the other along the tangent, then if the impetus is increased only in the latter direction and the moving body recedes from the center, the orbit will cease to be circular. But if the forces are increased equally in both directions, the motion will remain circular but accelerated by conatus, which no more argues that the forces of receding from the axis are increased than are the forces of approach toward it. It must therefore be said that water twirled around in a bucket rises to the sides of the vessel because, as new forces are being applied in the direction of the tangent to every particle of water, in the same instant no new equal centripetal forces are being applied.[40] From which experiment it in no way follows that absolute circular motion is necessarily discerned by the forces of receding from the axis of motion. Further, how these terms 'corporeal forces' and 'conatus' are to be interpreted is more than sufficiently explained in what has been premissed.

§61. Just as a curve can be considered as consisting of an infinity of right lines, even if in truth it does not consist of them but because this hypothesis is useful in geometry, in the same way circular motion can be regarded as traced and arising from an infinity of rectilinear directions, which supposition is useful in the mechanical philosophy. But it is not to be affirmed on that account that it is impossible for a center of gravity of

---

[40] This begins Berkeley's response to Newton's famous "bucket argument" from the Scholium to the definitions to the *Principia*.

any body to exist succesively in single points of the periphery of a circle, taking no account of any rectilinear direction, either in the tangent or the radius.

§62. It must by no means be ommited that the motion of a stone in a sling, or of water in a twirled bucket cannot be called truly circular, according to the understanding of those who define the true location of bodies by the parts of absolute space; for it is strangely composed of the motions not only of the bucket or sling, but also of the daily motion of the earth about its axis, its monthly motion about the common cener of gravity of the earth and moon, and its annual motion about the sun. And because of this, each particle of the stone or the water describes a line far removed from circular. Nor in truth is it, as is believed by some, an axifugal conatus, since it does not respect some one axis in relation to absolute space, even supposing there to be such a space: and so I do not see how it can be called a single conatus, to which a truly circular motion corresponds as to its proper and adequate effect.

§63. No motion can be discerned or measured except by sensible things. Since therefore absolute space in no way affects the senses, it is necessarily quite useless for distinguishing motions. Beyond this, determination or direction is essential to motion, and this consists in relation. Therefore it is impossible that absolute motion should be conceived.

§64. Furthermore, since by the diversity of relative locations, the motion of the same body varies, and indeed any thing can be said to be moved in one respect, and at rest in another: for determining true motion and true rest, by which means ambiguity is eliminated and the mechanics of those philosophers who contemplate a wider system of things is furthered, it would suffice to take the relative space enclosed by the fixed stars, regarded as at rest, instead of absolute space. Indeed motion and rest defined by such a relative space can conveniently be applied in place of the absolutes, which cannot be discerned by any mark. For however forces may be impressed, whatever conatus there may be, we admit that motion is to be distinguished by actions exerted on bodies; but never will it follow that there is this absolute space, and location, and the change of this is true motion.

§65. The laws of motions and effects, and the theorems containing the calculations of the same for different figures of the paths, as well for accelerations and diverse directions, and for more or less resistant media, all these hold without the calculation of absolute motion. Just as it is plain from the fact that, according to the principles of those who introduce absolute motion, it cannot be known by any mark whether the entire frame of things is at rest or moved uniformly in a right line, it is evident that the absolute motion of no body can be known.

§66. From what has been said it is clear that in investigating the true nature of motion, it will be of greatest avail first, to distinguish between mathematical hypotheses and the nature of things; second, to beware of abstractions; third, to consider motion as something sensible, or at least imaginable, and to be content with relative measures. Which things, if we do them, will at once leave untouched all the famous theorems of the mechanical philosophy, through which the recesses of nature are opened up and the system of the world is subjected to human calculation, while the consideration of motion will be freed from a thousand minutiae, subtleties, and abstract ideas. And let what has been said on the nature of motion suffice.

§67. It remains for us to discuss the cause of the communication of motions. Many judge that an impressed force in a mobile body is the cause of motion in it. But yet, that they do not assign a known

cause of motion, and one distinct from body and motion, is evident
from what has been premissed. It is clear, moreover, that force is not
a certain and determinate thing, from the fact that the greatest men
hold many diverse, even contrary, opinions on it, but nevertheless retain
truth in their results. For Newton says that impressed force consists
solely in action, and is the action exerted on a body to change its state,
nor does it remain after the action.[41] Torricelli contends that a certain
accumulation or aggregate of impressed forces is received by percussion
in a mobile body, and that the same remains and constitues impetus.[42]
Borelli and others say nearly the same thing.[43] And in truth, although
Newton and Torricelli seem to disagree, nevertheless, each advances a
consistent account, and the matter is adequately explained by both. For
all forces attributed to bodies are as much mathematical hypotheses as
are attractive forces in the planets and the sun. Mathematical entities,
however, have no stable essence in the nature of things: they depend on
the notion of the definer: whence the same thing can be explained in
different ways.

§68. Let us agree that the new motion in a body struck is conserved, either by the innate force through which every body persists in
its state, either of motion or of uniform motion in a right line, or by
the impressed force received during percussion into the body struck and
there remaining; this will be the same so far as the facts, the difference

---

[41] In commenting upon his definition of 'impressed force' (Definition IV in
the *Principia*, Newton claims "This force consists in action only, and remains
no longer in the body when the action is over. For a body maintains every new
state it acquires, by its inertia only. But impressed forces are of different origins,
as from percussion, from pressure, from centripetal force." (*Principia*, 1: 2)

[42] Torricelli argues for this view in Lecture III of the *Lezioni Accademiche*.
See Torricelli (1715, 15-17).

[43] Borelli's opinion is expressed in Chapter VI of his *De Vi Percussionis*. See
Borelli (1667, 48)

existing only in the names. Similarly, where a striking mobile body loses motion and the struck body acquires motion, it makes little difference to dispute whether the acquired motion is numerically the same with the lost motion, for indeed this leads only to metaphysical and even verbal minutiae conerning identity.[44] And so it comes to the same thing whether we say that motion passes from the striking to the struck, or that in percussion motion is generated anew, but destroyed in the striking body. In each case it is understood that one body loses motion, the other acquires motion, and beyond that nothing.

§69. I would hardly deny that the mind which moves and contains this universal corporeal mass is the true efficient cause of motion, and is the same cause, properly and strictly speaking, of the communication of this motion. But in physical philosophy, causes and solutions of the phenomena should be sought in mechanical principles. Therefore a thing is explained physically not by assigning its truly active and incorporeal cause, but by demonstrating its connection with mechanical principles: one of which is this, "that action and reaction are always contrary and equal,"[45] from which as from a source and primary principle, those rules of the communication of motions are drawn, which have already been found out and demonstrated by the moderns, to the great benefit of the sciences.

§70. It would be enough for us, if we should hint that this principle could have been declared in another way. For if the true nature of things, rather than abstract mathematics, is regarded, it will seem more correct to say that in attraction or percussion the passivity [passio] of bodies

---

[44] This recalls Berkeley's dismissal of philosophical disputes over identity in the third of his *Three Dialogues*: "whether philosopers shall think fit to call a thing the *same* or no, is, I conceive, of small importance." (*Works*,2: 247)

[45] This is Newton's third law of motion.

rather than the action is equal on both sides. For example, a stone linked by a rope to a horse is as much drawn toward the horse, as the horse is to the stone: for a body in motion colliding with one at rest suffers the same change as the quiescent body. And as regards the real effect, the striker is as the struck, and the struck as the striker. But this change on both sides, both in the body of the horse and in the stone, both in the moving body and in the quiescent, is meer passivity [*passio*]. But it is not the case that there is any force, virtue, or corporeal action truly and properly causing such effects. The moving body is driven against the quiescent, but we speak in the active voice, saying that this one impells that one; nor is this absurd in mechanics, where mathematical ideas rather than the true nature of things are regarded.

§71. In physics sense and experience, which extend only to apparent effects, have their place; in mechanics the abstract notions of the mathematicians are admitted. In first philosophy or metaphysics incorporeal things are concerned, such as causes, truth, and the existence of things. The physicist contemplates the series or succession of sensible things, observing by which laws they are connected, and in what order, what precedes as a cause, and what follows as an effect. And in this way we say that a moving body is the cause of motion in another, or that it impresses motion on it, pulls it or impells it. In this sense second corporeal causes should be understood, no account being taken of the actual seat of forces, or active powers, or of the real cause in which they are.[46] Moreover, besides body, figure, and motion, even the mechanical

---

[46] So-called second causes are the finite bodies in the world which depend upon God (the first cause) for their existence. Second causes were traditionally interpreted as the means through which God acts in the world, since all events in the world are ultimately the result of God's will. The occasionalist doctrine held that second causes are actually not causes at all, but merely the "occasion" for God's action. Berkeley's doctrine here is thus very similar to the occasionalist thesis.

principles or the primary axioms of mechanical science can be called causes, being regarded as the causes of the consequences.

§72. Only by meditation and reasoning can truly active causes be brought to light from out of the enveloping darkness, and to some extent known. But to treat of them is the concern of first philosophy or metaphysics. And if to each science its province were alloted, its limits assigned, and the principles and objects which belong to it accurately distinguished, we could treat each with greater ease and perspicuity.

FINIS

The Analyst

Editor's Introduction

The Analyst is a work which merits serious attention from students of Berkeley's philosophy, but it has not generally been the focus of sustained scholarly consideration. The reasons for this relative neglect are no doubt complex, but part of the problem seems to be that *The Analyst* is a critique of seventeenth and eighteenth-century mathematical theories which are not generally well-known to contemporary historians of philosophy. Nevertheless, once the fundamental mathematical issues are understood *The Analyst* can be read as an important statement of Berkeley's conception of rigorous mathematical demonstration, and a close reading of the text becomes an essential part of our understanding of Berkeley's account of science and mathematics.

Philosophers' attention to Berkeley has tended to focus on his idealistic metaphysics to the virtual exclusion of all other aspects of his thought. The notion that the "real Berkeley" is to be found in the arguments against materialism tends to downplay the significance of his mathematical writings, since his account of mathematics is independent of his views on materialism. But if we take a broader view of Berkeley's project we can take his concern with mathematics seriously and seek to understand his philosophy of mathematics on its own terms. What Berkeley has to say about the calculus is quite interesting enough to warrant our attention, and a full understanding of Berkeley's philosophical position is impossible without paying attention to *The Analyst*.

But *The Analyst* is more than a contribution to eighteenth-century philosophy of mathematics, for it is also an attack on an unnamed "infidel mathematician" and a serious piece of Christian apologetics. The theological aim of *The Analyst* is to show that accepted mathematical theories contain mysteries and fallacious reasoning, so that freethinkers who deride Christianity for its alleged incomprehensibility cannot also accept the calculus of Newton and Leibniz. This theological strategy is obviously parasitic upon Berkeley's mathematical argumentation, so it makes sense to begin by setting the mathematical background to *The Analyst* and then proceed to an account of its theological context. It should be stressed, however, that the cogency of Berkeley's mathematical argumentation does not depend upon his theological commitments. That is to say, we can accept his critique of the calculus without sharing his concern with freethinking and atheism.

1. THE MATHEMATICAL BACKGROUND

Berkeley's target in *The Analyst* is the collection of mathematical techniques which he styles "the modern Analysis," but which are known today as the calculus. The calculus was developed in the second half of the seventeenth century as a method for solving important general problems in analytic geometry.[1] For example, one of these central problems was the following: given an equation which determines the analytic representation of a curve in Cartesian co-ordinates, find the tangent to the curve at an arbitrary point. Similarly, such problems as finding the area enclosed by a curve (known as the problem of "quadrature" in the parlance of the period) or the determination of arc-lengths for a given curve (known as "rectification") are all standard fare in the calculus.

1.1 THE CLASSICAL STANDARD OF RIGOR

The methods of proof deriving from classical sources were generally insufficient to solve these problems in a completely general form, but they provided an unchallenged model of rigorous demonstration and it is worthwhile to sketch briefly how mathematicians of the classical

---

[1] Writings on the history of the calculus are voluminous. A dated standard work is Boyer ([1949] 1959), which can be supplemented by the accounts in Boyer ([1968] 1989), Baron (1969), Edwards (1979), and the papers in Grattan-Guinness (1980). Useful anthologies of primary sources are Struik ([1969] 1986) and Fauvel and Gray (1987).

period approached such problems as quadrature. Classical authors conceived of mathematics as the science of quantity in general, but saw it as consisting of two distinct sciences (arithmetic and geometry), distinguished from one another by the fact that geometry studied continuous quantity ("magnitude") while arithmetic dealt with discrete quantity ("multitude" or "number"). Problems of tangency or quadrature are, on this account, problems in geometry and must be addressed within the framework of classical geometry.

Within the genus of magnitude there are several species. These were taken to be the fundamentally different kinds of objects which can be studied by geometric methods, including lines (both straight and curved), angles, arcs, surfaces, and solids. It is significant that ratios can be formed by the objects within a species, but it is impossible to make ratio comparisons across species. Thus, the ratio between two right lines or two angles can be formed, but there is no way to compare a surface to a line or an angle to a solid.

It is in the theory of ratios of magnitudes that the finitistic nature of classical geometry is most clearly manifest. The codification of the principles which govern magnitudes and their ratios is Book V of Euclid's *Elements*, where the general theory of proportions is developed. The key concepts are contained in the third through sixth definitions of Book V, which read as follows:

> 3. A *ratio* is a sort of relation in respect of size between two magnitudes of the same kind.
> 4. Magnitudes are said to *have a ratio* to one another which are capable, when multiplied of exceeding one another.
> 5. Magnitudes are said to *be in the same ratio*, the first to the second and the third to the fourth, when, if any equimultiples whatever be taken of the first and third, and any equimultiples whatever of the second and fourth, the former equimultiples alike exceed, are alike equal to, or alike fall short of, the latter equimultiples respectively taken in corresponding order.
> 6. Let magnitudes which have the same ratio be called *proportional*. (Euclid [1925], 1956 2: 114)

The significance of these definitions is that they provide the means for comparing magnitudes within each species by the formation of ratios, and then comparing ratios across species of magnitudes by constructing proportions. The finitistic character of the classical theory should be apparent, especially when it is understood that the multiplications referred to in Definitions 4 and 5 are finite multiplications. To compare two magnitudes α and β in a ratio $\alpha : \beta$, it is necessary that continued multiplication of one make it exceed the other. This explicitly bars division by zero (or its geometric equivalents), and it prevents ratios from being formed across species because there is no multiplication of a line which will allow it to exceed an angle or surface. But proportions can be constructed from ratios whenever the criterion in Definition 5 is satisfied, so it makes sense to ask whether the ratio between two given lines is the same as that between two given spheres, even though a line and sphere cannot be compared directly with one another. This theory of proportion was put to use throughout classical geometry, with the standard form of a problem being that of finding the ratios and proportions between various geometric magnitudes. Thus, a quadrature of a figure enclosed by a curve would be stated as the problem of finding a square equal in area to the figure.

The principal style of proof licensed by the classical conception of magnitudes is the so-called "method of exhaustion" in which an unknown quantity (or the ratio between two unknowns) is determined by considering sequences of known quantities which can be made to differ from the unknown by an arbitrarily small amount. This technique set the standard of rigor for the investigation of quadratures, and it merits a brief survey before we explore some of the new methods critiqued by Berkeley

The foundation of the method of exhaustion is Proposition 1 of Book X of Euclid's *Elements*. This proposition follows immediately from Definition 4 of Book V, and its use is essential in the course of an exhaustion proof when a sequence of approximations is shown to differ from a given magnitude by less than any assigned amount. The proposition reads:

Two unequal magnitudes being set out, if from the greater there be subtracted a magnitude greater than

its half, and from that which is left a magnitude greater
than its half, and if this process be repreated continually, there will be left some magnitude which will be
less than the lesser magnitude set out. (Euclid [1925]
1956, 3: 14)

The general procedure in an exhaustion proof is to begin with upper and
lower bounds for an unknown magnitude and then to provide a method
for systematically improving these bounds. In the case of an exhaustion
proof to determine the quadrature of a figure, the initial bounds will be
given in the form of inscribed and circumscribed figures. Then a method
for improving these bounds must be exhibited, typically by inscribing
and circumscribing two new figures which reduce the remainder between
the bounds and the unknown by more than half. If this method can
be iterated, it generates a sequence of improved approximations which
[by Euclid (X, 1)] will differ from the unknown by less than any given
magnitude.

As an example of this procedure, the area of a circle can be bounded
above and below by inscribing and circumscribing squares. If we then
double the number of sides in our approximating figures to form inscribing and circumscribing octagons, we can reduce the difference between
the area of the circle and the areas of the approximations by more than
half. Moreover, by continuing to double the number of sides in our approximations we can form two sequences, such that successive terms of
each sequence reduce the remainder by more than half. Thus, by Euclid
(X, 1), the difference between the area of the circle and the area of the
inscribed and circumscribed polygons can be made as small as desired.
When such a "compression" of the value of an unknown between two
sequences of known quantities is attained, the classical exhaustion proof
is rounded off by a double *reductio ad absurdum* which shows that the
unknown value can be neither greater nor less than a specified amount.

Two points should be stressed here. First, there is no need to consider infinitely small quantities in the course of an exhaustion proof.
Throughout the course of the proof we make reference only to the finite
differences between finite magnitudes, and the process of determining
the area of the unknown requires only a finite number of steps. This
avoidance of infinitary considerations is rooted in the definitions from

Book V of the *Elements*, and it is on this account that the mathematical analysis of the seventeenth and eighteenth centuries differs most
substantially from the Greek model. The second point that should be
clear is that a fully executed exhaustion proof is a very cumbersome
chain of argument which cannot easily be generalized to cover a variety
of cases. The proof requires the specification of both a value toward
which the sequences of approximations tend and a method for generating sequences of approximating figures. In general it is difficult to
achieve these results, and the required double *reductio ad absurdum* can
make the proofs of even the most elementary theorems unmanageably
long and intricate.

   These two features of the method of exhaustion were widely acknowledged by mathematicians of the seventeenth century, who agreed
that exhaustion proofs were paradigmatically rigorous but complained
that the technique was cumbersome and lacked the generality to cover
any but the most elementary cases. This frustration with the method
eventually led to the development of infinitesimal mathematics – an
episode to which we must now turn our attention.

1.2 INFINITESIMAL CALCULUS

One way to abandon the finitistic standpoint of classical mathematics
is to regard geometric objects as infinite collections of infinitely small
elements; thus, a line could be taken as an infinte collection of points,
a surface as an infinite collection of lines, etc. In this procedure the
classical division between species of magnitude is broken down: whereas
the classical model treats the separate species as fundamentally distinct,
this account treats an object from one species of magnitude as composed
of an infinity of objects from a species of lower dimension. Such an approach became popular in the seventeenth century and was known as
the "method of indivisibles." Its first exposition was in Bonaventura
Cavalieri's *Geometria indivisibilibus continuorum nova quadam ratione*

promota (1635), and it quickly became the preferred approach to problems of quadrature.[2] The method plays upon the intuition that we can find the area of a figure by considering the lines it contains (known as the "indivisibles of the figure"). Cavalieri was cautious about claiming that these indivisibles actually composed the figure, but the progressive mathematicians of the seventeenth century represented geometric problems analytically in Cartesian co-ordinates and solved them by determining the relationship between the infinite sums of indivisibles which they took to compose the figures.

As an example, take John Wallis's quadrature of the cubic parabola in his *Arithmetica Infinitorum* (1656). He begins with arithmetical results, observing that

$$\frac{0+1}{1+1} = \frac{1}{2} = \frac{2}{4} = \frac{1}{4} + \frac{1}{4}$$

$$\frac{0+1+8}{8+8+8} = \frac{9}{24} = \frac{3}{8} = \frac{1}{4} + \frac{1}{8}$$

$$\frac{0+1+8+27}{27+27+27+27} = \frac{36}{108} = \frac{4}{24} = \frac{1}{4} + \frac{1}{12}$$

$$\frac{0+1+8+27+64}{64+64+64+64+64} = \frac{100}{320} = \frac{5}{16} = \frac{1}{4} + \frac{1}{16}$$

From these initial cases, Wallis concludes "by induction" that as the number of terms in the sums increases, the ratio approaches arbitrarily near to $\frac{1}{4}$. Then, in Theorem 41, he declares:

> If an infinite series of quantities which are the cubes of a series of continuously increasing numbers in arithmetic progression, beginning with 0, is divided by the sum of numbers all equal to the highest and equal in number, then we obtain $\frac{1}{4}$. This follows from the preceding reasoning.[3]

---

[2] See Andersen (1985), Giusti (1980), and Jesseph (1989) for an account of Cavalieri's method and its reception.

[3] In this and the following quotation I have used the translation in Struik ([1969] 1986, 245-246) The Latin can be found in Wallis (1693-99, 1: 382-383).

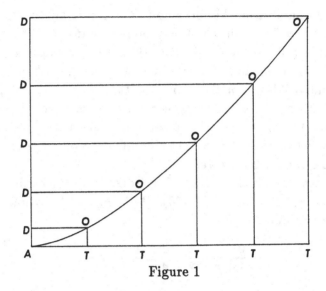

Figure 1

Given this result, Wallis turns to the quadrature of the cubic parabola, treating it as an infinite sum of lines forming a series of cubic quantities:

> The complement AOT of half the area of the cubic parabola therefore is to the parallelogram TD over the same base and altitude as 1 to 4. Indeed, let AOD [in Figure 1] be the area of half the parabola AD (its diameter AD, and the corresponding ordinates DO, DO, etc.) and let AOT be its complement. Since the lines DO, DO, etc., or their equals AT, AT, etc. are the cube roots of AD, AD, . . ., or their equals TO, TO, . . . these TO, TO, etc. will be the cubes of the lines AT, AT, ... The whole figure AOT therefore (consisting of the infinite number of lines TO, TO, etc., which are the cubes of the arithmetically progressing lines AT, AT, . . .) will be to the parallelogram ATD (consisting of just as many lines, all equal to the greatest TO), as 1 to 4, according to our previous theorem. And the half-segment AOD of the parabola (the residuum of the parallelogram) is to the parallelogram itself as 3 to 4. (Struik [1969] 1986, 246; Wallis 1693-99, 1: 383.)

What is interesting here is Wallis's conception of surfaces as composed of indivisibles and his application of algebraic and arithmetical results on the summation of infinite series to problems in geometry. But note also that the method departs from the classical approach to geometry because it fails to observe the classical distinction between discrete and continuous magnitudes. By treating a continuous geometric figure as composed of sums of discrete points or lines, the method of indivisibles simply ignores the classical distinction. Moreover, the method's dependence upon infinite summations and the introduction of infinitely small magnitudes is a complete departure from the finitistic viewpoint codified in the classical theory of ratios.

The differential calculus of Leibniz and his followers goes beyond the method of indivisibles and uses infinitesimal considerations to develop a general theory which can solve problems of tangency and quadrature.[4] The fundamental concept here is that of the difference of a variable; given a variable x, the difference dx is taken to be the difference between two values of x which are infinitely close to one another. An alternative account of differences characterizes them as magnitudes which stand in the same ratio to a finite magnitude that any finite magnitude stands to infinity. The intent of this definition is to capture the two properties of the infinitesimal: being greater than zero but less than any real number. As ratios of a finite to an infinite, differences are greater than zero (since they are ratios of positive magnitudes), but they remain less than any finite magnitude (because any real number can be expressed as the ratio of finite magnitudes). Obviously, this account departs substantially from the classical theory of ratios, where two magnitudes can have a ratio only if they are capable of exceeding one another by a finite multiplication.

The Leibnizian presentation of the calculus also introduces differences of differences (so-called "second differences") which are still positive magnitudes, but are infinitely small in comparison with a difference of the first order. The expressions 'ddx' or 'dx^2' are introduced for these second differences, which were commonly taken to denote the product of two differences of the first order. Multiplication of a finite magnitude by a first difference will yield an infinitely small magnitude, while the

---

[4] See Bos (1974) and the anthology by Heinekamp (1986) for specific presentations of the Leibnizian calculus.

multiplication of two differences of the first order yields a magnitude infinitely smaller than either of its factors, but still greater than zero. Naturally, the theory permits continued multiplication of infinitesimals to generate infinitesimals of all orders.

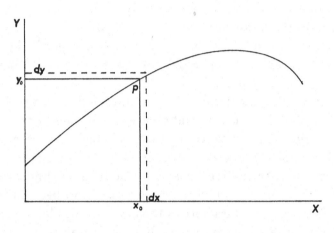

Figure 2

Solutions to problems of tangency arise quite naturally out of this conception of infinitesimal differences. Given the curve $\alpha\beta$ as in Figure 2, the tangent at point p with the co-ordinates (x_0, y_0) can be constructed by taking the differentials dy and dx along the axes x and y; these will form an infinitesimal rectangle whose diagonal will correspond with the tangent at p, and we can express the tangent as $\frac{dy}{dx}$. Alternatively, the curve can be treated as a polygon with an infinite number of infinitely small sides, and the tangent can be taken as a line coincident with the side at p.

The great strength of the calculus lies in its generality, and the differential calculus reduces the problem of tangency to a simple algorithm (known as differentiation) for curves which can be represented analytically by an equation. For example, suppose we take the equation

$$y = x^3 + 5x^2 - 4x + 1 \qquad\qquad [1]$$

Using the differences dy and dx, we investigate the infinitesimal increment of the curve and obtain

$$(y + dy) = (x + dx)^3 + 5(x + dx)^2 - 4(x + dx) + 1 \qquad [2]$$

Expanding equation [2] yields

$$(y + dy) = x^3 + 3x^2 dx + 3x dx^2 + dx^3 + 5x^2 + 10x dx$$
$$+ 5dx^2 - 4x - 4dx + 1 \qquad [3]$$

Equation [3] represents [1] augmented by the increments dy and dx, but the increment itself can be obtained by subtracting [1] from [3]. Thus we get

$$dy = 3x^2 dx + 3x dx^2 + dx^3 + 10x dx + 5dx^2 - 4dx \qquad [4]$$

Simplifying [4] by dividing through by dx will give the ratio of the two increments dy and dx, which is expressed as

$$\frac{dy}{dx} = 3x^2 + 3x dx + dx^2 + 10x + 5dx - 4 \qquad [5]$$

But because dx is infinitely small when compared to x, we can drop the terms containing it from the right side of equation [5] and retain only terms in x, which results in

$$\frac{dy}{dx} = 3x^2 + 10x - 4 \qquad [6]$$

and this equation gives the tangent at any point on the curve.

Even more strikingly, the calculus can relate the problems of quadrature and tangency by showing that computing a quadrature and finding a tangent are inverse operations. This inverse relationship can be brought out by treating the area under the curve $\alpha\beta$ on the interval $[0, a]$ as the sum of the ordinates y over the interval (Figure 3). Using the familiar notation for integration, this area can be written "$\int_0^a y dx$."

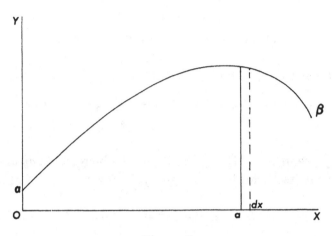

Figure 3

Alternatively, the integral can be treated as the sum of the infinitely narrow rectangles formed by the ordinate y and the differential dx.

To show that quadrature and tangency are inverse operations we introduce a new curve defined as the integral with a variable upper bound x; using the modern notation of functions, this curve will be given by the equation $G(x)$, where

$$G(x) = \int_0^x y\,dx \qquad [7]$$

Then the tangent to this curve $G(x)$ at an arbitrary point with abscissa x_0 can be written as

$$\frac{G(x_0 + dx) - G(x_0)}{dx} \qquad [8]$$

But $G(x_0 + dx)$ is the area between 0 and $x_0 + dx$, and $G(x_0)$ represents the area between 0 and x_0. Therefore, $G(x_0 + dx) - G(x_0)$ gives the area between x_0 and $x_0 + dx$. Then

$$G(x_0 + dx) - G(x_0) = y\,dx \qquad [9]$$

Dividing through by dx in equation [9] gives

$$\frac{G(x_0 + dx) - G(x_0)}{dx} = \frac{y\,dx}{dx} = y \qquad [10]$$

Thus, the instantaneous rate of change of the function which represents the area under the curve $\alpha\beta$ (or, the tangent to the curve which represents the area under the original curve $\alpha\beta$) is the same as the ordinate of the curve $\alpha\beta$. Using the analytic representations of the curves, we can express this inverse relationship by saying that the equation which expresses the area under the curve $\alpha\beta$ on the interval $[0, x]$ is one that, when differentiated, yields the ordinate of the curve $\alpha\beta$ at x. This fundamental relationship between differentiation and integration, combined with the algorithmic character of the infinitesimal calculus, allowed much easier and more general solutions to problems of tangency, quadrature, and rectification than were available within the confines of classical methods.

1.3 THE NEWTONIAN CALCULUS OF FLUXIONS

Newton's presentation of the calculus relies upon a kinematic conception of geometric magnitudes in which lines, angles, surfaces, and solids are taken to be produced by continuous motion. This account is not a Newtonian innovation: it dominates Isaac Barrow's *Geometrical Lectures*[5] and has classical antecedents in the treatment of certain "mechanical" curves which are defined in terms of the motion of points and lines. In the Introduction to his 1704 treatise *On the Quadrature of Curves*, Newton explicitly opposes this theory to the Leibnizian conception of magnitudes as composed of infinitesimal parts:

> Mathematical quantities I here consider not as consisting of least possible parts, but as described by a continuous motion. Lines are described and by describing are generated, not through the apposition of parts but through the continuous motion of points; surface-areas are through the motion of lines, solids through the motion of surface-areas, angles through the rotation of sides, times through a continuous flux, and the like in other cases. These geneses take place in the reality of

---

[5] See Lectures 1 and 2 of the *Lectiones Geometricæ* in Barrow (1860, 2: 159-185) where Barrow considers the "generation of magnitudes."

physical nature and are daily witnessed in the motion
of bodies. And in much this manner the ancients, by
'drawing' mobile straight lines into the length of sta-
tionary ones, taught the genesis of rectangles. (*Papers*,
8: 123)

In Newtonian terminology a flowing quantity is called a *fluent* and the
velocity with which it is produced is called its *fluxion*. Notationally,
fluents are represented by variables x and y, while their fluxions are
dotted letters $\dot{x}$ and $\dot{y}$. Higher-order fluxions can be generated by treating a fluxion as itself a flowing quantity, so that $\ddot{x}$ and $\ddot{y}$ would indicate
the fluxions of the fluxions of the original fluents x and y.

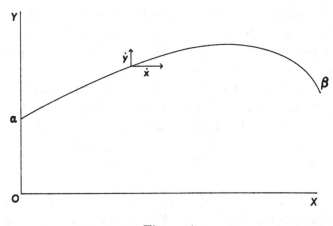

Figure 4

A curve such as $\alpha\beta$ in Figure 4 can be seen as produced by the motion of a point in the Cartesian plane, and its fluxion will consist of two
components, $\dot{x}$ and $\dot{y}$, parallel to the axes OX and OY. The fundamental
problems of the calculus can now be phrased in terms of fluxions and
fluents. Tangency problems will become problems of finding the fluxions $\dot{x}$ and $\dot{y}$ when given an equation which describes the relationship
between the fluents x and y. A quadrature will be an inverse problem,
that of determining the fluents when the fluxions are given.

In the solution of these problems Newton developed two devices. The first is his doctrine of moments, and the second his theory of prime and ultimate ratios. The moment of a fluent is defined as its "momentaneous synchronal increment," or the amount by which a fluent is increased in an "indefinitely small" period of time. These periods of time are represented by the symbol 'o' and the moment of the fluent x will be $o\dot{x}$. Thus, the fluents x and y will be augmented by their moments in an indefinitely small period of time to become $x + o\dot{x}$ and $y + o\dot{y}$.

The theory of prime and ultimate ratios is closely tied to the kinematic conception of magnitudes and involves the consideration of ratios between magnitudes as they are generated by motion. The prime ratios of nascent magnitudes are those which hold as the magnitudes are just beginning to be generated, while the ultimate ratios of evanescent magnitudes are ratios holding between magnitudes which are diminished to nothing and vanish. Newton gave several expositions of this theory, with one fairly straightforward account coming in the "Introduction" to the *Quadrature of Curves*:

Fluxions are very closely near as the augments of their fluents begotten in the very smallest equal particles of time: to speak accurately, indeed, they are in the first ratio of the nascent augments, but they can, however, be expressed by any lines whatever which are proportional to them. If, for instance, the areas (ABC), ($ABDG$) [in Figure 5] be described by the ordinates BC, BD advancing upon the base AB with uniform motion, the fluxions of these areas will be to one another as the describing ordinates BC and BD, and can be expressed by those ordinates, for the reason that those ordinates are as the nascent augments of the areas. Let the ordinate BC advance from its place BC into any new place bc, complete the parallelogram $BCEb$, and draw the straight line $VCTH$ to touch the curve at C and meet bc and BA extended in T and V: then the augments of the abscissa AB, ordinate BC and the curve-arc $\overset{\frown}{AC}$ now begotten will be Bb, Ec and $\overset{\frown}{Cc}$, and in the first ratio of these nascent augments are the sides

of the triangle CET; consequently the fluxions of AB,
BC, and AC are as the sides CE, ET and CT of that
triangle CET, and can be expressed by means of these
same sides or, what is the same, by the sides of the
triangle VBC similar to it. (*Papers*, 8: 125)

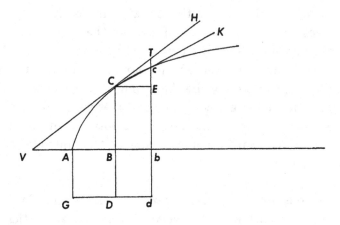

Figure 5

To see how the doctrines of moments and ultimate ratios work in
the calculus of fluxions, consider the problem of determining the fluxion
of a fluent, as set forth in the *Quadrature of Curves*. Newton uses the
equation

$$x^3 - xy^2 + a^2 z - b^3 = 0 \qquad [11]$$

with fluents x, y and constants a, b. He begins by taking the moments
$o\dot{x}, o\dot{y}, o\dot{z}$ of the flowing quantities and substituting them as increments
into equation [11]. This yields

$$(x + o\dot{x})^3 - (x + o\dot{x})(y + o\dot{y})^2 + a^2(z + o\dot{z}) - b^3 = 0 \qquad [12]$$

By expansion, [12] becomes

$$x^3 + 3x^2 o\dot{x} + 3xo^2\dot{x}^2 + o^3\dot{x}^3 - xy^2 - o\dot{x}y^2 - 2xo\dot{y}y -$$
$$2\dot{x}o^2\dot{y}y - xo^2\dot{y}^2 - \dot{x}o^3\dot{y}^2 + a^2 z + a^2 o\dot{z} - b^3 = 0 \qquad [13]$$

Equation [13] represents equation [11] plus the increments $\dot{x}o$, $\dot{y}o$, and $\dot{z}o$. The difference between [13] and [11] will thus give the increment of the original equation. Subtracting [11] from [13] yields:

$$3x^2o\dot{x}+3xo^2\dot{x}^2 + o^3\dot{x}^3 - o\dot{x}y^2 - 2xo\dot{y}y- \\ 2\dot{x}o^2\dot{y}y - xo^2\dot{y}^2 - \dot{x}o^3\dot{y}^2 + a^2o\dot{z} = 0 \quad\quad [14]$$

Dividing [14] by o, we obtain:

$$3x^2\dot{x} + 3xo\dot{x}^2 + o^2\dot{x}^3 - \dot{x}y^2 - 2x\dot{y}y - 2\dot{x}o\dot{y}y - xo\dot{y}^2 - \dot{x}o^2\dot{y}^2 + a^2\dot{z} = 0 \quad [15]$$

If we now "let the quantity o be lessened infinitely" and neglect the "evanescent terms" which contain o as a factor, we obtain

$$3x^2\dot{x} - \dot{x}y^2 - 2x\dot{y}y + a^2\dot{z} = 0 \quad\quad [16]$$

as the equation which determines the fluxion of the original equation. This procedure can then be extended to an algorithm exactly analogous to the procedure of differentiation in the Leibnizian calculus.

The second problem (that of determining a fluent, given the fluxion) proceeds inversely and can be used to determine quadratures in essentially the same way that the Leibnizian calculus takes integration as the inverse of differentiation. The quadrature of the curve $\alpha\beta$ in Figure 6 will yield the area $\alpha\beta\gamma\delta$ bounded by the curve, the abscissa and the two ordinates. According to the kinematic conception of magnitudes, the area is swept out by the ordinate as it moves from $\delta\alpha$ to $\gamma\beta$. The equation for the curve gives us the value of the ordinate at any point in its transversal of the abscissa, and the fluxion of the area will be the ordinate (since this fluxion is the rate of the increase of the area). Thus, if we are given the equation for the curve we are thereby given the equation for the fluxion of the area, and the quadrature of the curve can be obtained by finding an equation whose fluxion is the equation of the curve $\alpha\beta$.

As we will see, Berkeley does not challenge the important results delivered by the method of fluxions or the *calculus differentialis*, but he is concerned with important foundational issues. Of critical importance here is the role of "evanescent magnitudes" or discarded infinitesimal differences. The main point of Berkeley's attack is that these procedures

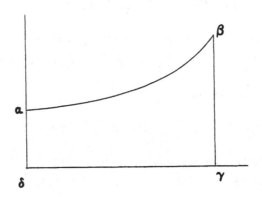

Figure 6

offend against principles of sound reasoning by introducing contradictory assumptions into a demonstration. The contradiction arises when a magnitude is treated as positive when we wish to divide by it (as when equations [15] and [5] are obtained through division by o or dx) and zero when we wish to cancel terms containing it (as in equations [6] and [16]).

2. THE THEOLOGICAL BACKGROUND

The title page of *The Analyst* attributes the work to "the Author of *The Minute Philosopher*," and it is a continuation of the battle against free-thinking which Berkeley undertook in his 1732 work *Alciphron; or, The Minute Philosopher*. One of Berkeley's principal targets in *Alciphron* is the freethinkers' doctrine that all religious conviction must ultimately be based upon reason rather than faith.[6] This doctrine was expressed by numerous writers of the period, but most clearly by Anthony Collins,

---

[6] Alternatively, one could characterize Berkeley's target as the doctrine that a natural theology based on reason is superior to a revealed theology based on faith. Berkeley is not, of course, opposed to reason, or even to the project of a natural theology. But his commitments to Anglican orthodoxy demand that faith and mysteries be given their due.

John Toland, and Matthew Tindal. Toland's 1696 *Christianity not mysterious; or, A treatise shewing that there is nothing in the gospel contrary to reason, nor above it; and that no Christian doctrine can be properly call'd a mystery* set the tone for this genre of theological literature; Toland argued that the truths of revelation, although not actually discovered by reason, are fully consistent with it and could have been attained by unaided reason. Collin's 1713 *Discourse of free-thinking* went farther and denied any role for revelation in true religion, insisting that superstition and tradition had burdened religion with many incoherent doctrines which were the source of error and dispute. Tindal's *Christianity as old as the creation, or The gospel a republication of the religion of nature* (1730) continued this line of thought and reduced Christianity to a set of doctrines which required neither Christ nor a God distinct from nature itself.

Berkeley's opposition to freethinking was longstanding,[7] but became much more active during his stay in Newport (1729-31), where he seems to have encountered numerous heterodox opinions among the colonial population. His chief complaint against freethinking is that it inexorably leads to atheism. In the first Dialogue of *Alciphron*, the "minute philosopher" and freethinker Alciphron depicts his progress toward atheism, which starts with an attempt to reconcile conflict among Christian sects and ends with the rejection of all religion:

> [H]aving observed several sects and subdivisions of sects espousing very different and contrary opinions, and yet all professing Christianity, I rejected those points wherein they differed, retaining only that which was agreed to by all, and so became a Latitudinarian. Having afterwards, upon a more enlarged view of things, perceived that Christians, Jews, and Mahometans had each their different systems of faith, agreeing ony in the belief of one God, I became a Deist. Lastly, extending my view to all the other various nations which inhabit

---

[7] In a 1713 series of essays (including one entitled "Minute Philosophers") in Sir Richard Steele's *The Guardian*, Berkeley attacked freethinkiners and made a case for revealed religion which is similar to that found in the *Alciphron*. See (*Works*, 7: 179-228).

> this globe, and finding they agreed in no one point of
> faith, but differed one from another, as well as from
> the forementioned sects, even in the notion of a God,
> in which there is as great diversity as in the methods
> of worship, I thereupon became an atheist; it being my
> opinion that a man of courage and sense should follow
> his argument wherever it leads him, and that nothing
> is more ridiculous that to be a free-thinker by halves.
> (*Works*, 3: 43-44)

Berkeley's strategy in *Alciphron* is to deny that the free-thinkers have
a monopoly on rationality. He aims to defend Anglican Christianity by
arguing that the freethinkers are victims of the same prejudice, irra-
tionality, and unsound reasoning which they purport to find in Christianity. This strategy is central to *The Analyst* as well, and is indicated
in the full title of the work, which purports to examine "whether the
Object, Principles, and Inferences of the Modern Analysis are more distinctly conceived, or more evidently deduced, than religious Mysteries
and Points of Faith."

The *Analyst* is addressed to an unnamed "infidel mathematician,"
who may have been Edmund Halley (1656-1742), although it is difficult
to be certain of this point. It seems clear that Berkeley had a specific
person in mind when he wrote *The Analyst*; he declares as much in §7
of his *Defense of Free-Thinking in Mathematics*:

> Whether there are such infidels, I submit to the judge-
> ment of the reader. For my own part I make no doubt
> of it, having seen some shrewd signs thereof myself,
> and having been very credibly informed thereof by oth-
> ers. . . . [The late Mr. Addison] assured me that the
> infidelity of a certain noted mathematician, still living,
> was one principal reason assigned by a witty man of
> those times for his being an infidel. Not that I imag-
> ine geometry disposeth men to infidelity; but that, from
> other causes, such as presumption, ignorance, or vanity,
> like other men geometricians also become infidels, and

that the supposed light and evidence of their science
gains credit to their infidelity. (*Works*, 4: 112)

Joseph Stock, an early but unreliable biographer of Berkeley, is our
source for identifying Halley as the "infidel mathematician" whose conduct was reported by Addison. Stock writes:

> The occasion [of *The Analyst*] was this: Mr. Addi-
> son had given the Bishop an account of their common
> friend Dr. Garth's behaviour in his last illness, which
> was equally unpleasing to both those excellent advo-
> cates for revealed religion. For when Mr. Addison
> went to see the Doctor, and began to discourse with
> him seriously about preparing for his approaching dis-
> solution, the other made answer, "Surely, Addison, I
> have good reason not to believe those trifles, since my
> friend Dr. Halley who has dealt so much in demonstra-
> tion, has assured me, that the doctrines of Christianity
> are incomprehensible, and the religion itself an impos-
> ture." The Bishop therefore took arms against this re-
> doubtable dealer in demonstration, and addressed *The
> Analyst* to him, with a view of shewing, that myster-
> ies in faith were unjustly objected to by mathemati-
> cians, who admitted much greater mysteries, and even
> falshoods [*sic*] in science, of which he endeavoured to
> prove that the doctrine of fluxions furnished an emi-
> nent example. Such an attack upon what had hitherto
> been looked upon as impregnable, produced a number
> of warm answers, to which the Bishop replied once or
> twice. (Stock [1776] 1989, 1: 29-30)

The difficulty with this account is that Samuel Garth died in January
of 1719, more than fifteen years before the publication of *The Analyst*
and during a period when Berkeley himself was in Italy. Addison died
in June of 1719 (while Berkeley was still in Italy), so he could only have
informed Berkeley of the incident by letter, but no such letter survives
in the scanty collection of Berkeley's correspondence from this period.

Ultimately, the identity of the "infidel mathematician" is of no great consequence in interpreting *The Analyst*, so we can leave the matter here. What is important is that Berkeley saw the enemies of revealed religion as taking mathematics as a paradigm of sound reasoning and argument, and then deriding religion because it fails to live up to the epistemological standard set by mathematics. The core of Berkeley's theological strategy in *The Analyst* will thus be to show that the calculus is no less mysterious than Christianity. Note, however, that Berkeley sees mystery as admissible (indeed, essential) in religion, while his criteria for mathematical and scientific rigor demand that the "object" of a genuine science be clearly conceived. He thus sees a fundamental difference between science and theology: theological matters contain mysteries which are beyond (but not contrary to) human reason, while science must deal only with things that are clearly conceived and evident to reason. This distinction is brought out nicely Query 62 at the end of *The Analyst*: "Whether Mysteries may not with better right be allowed of in Divine Faith than in Humane Science?"

3. BERKELEY'S CASE AGAINST THE CALCULUS

Berkeley's challenge to the calculus is addressed to a British audience, so he deals primarily with the Newtonian method of fluxions. Although he critiques the Leibnizian *calculus differentialis* as well, the polemics against the Continental procedures are more of an afterthought and appear only once Berkeley is satisfied that he has discredited the Newtonian doctrine. His basic claim against the calculus is that it is an unrigorous method. Mathematical rigor is a notoriously difficult concept to articulate, but we can usefully distinguish two respects in which a mathematical procedure might well be challenged, characterizing these as metaphysical and logical criteria of rigor. In the first place, a (putative) demonstration might make reference to certain kinds of objects which are thought to be conceptually or metaphysically problematic, thereby violating a metaphysical criterion of rigor. Constructivist objections to the use of infinitary proof techniques are typically of this sort, since the constructivist argues that we have no concept of an infinite totality and thus cannot reason about it. Berkeley intends just such an attack when he contrasts the object and principles of the calculus with religious mysteries. A second respect in which a mathematical procedure might be branded unrigorous is its use of inferences which are invalid or mistaken. Such criticisms invoke a logical criterion of rigor and are common whenever mathematical results are called into question. Berkeley indicates such a critique of the calculus when he asks whether the inferences of "the Modern Analysis" are as evidently demonstrated as points of faith. It is worthwhile to set out Berkeley's main arguments against the calculus to show how they fall into these two distinct categories.

The "metaphysical" critique of the calculus in *The Analyst* amounts to little more than Berkeley's claim that moments, fluxions, and infinitesimal differences are inconceivable. Moreover, his case for the inconceivability of the objects of the calculus is based largely on a first-person report of his own inability to imagine such things. He indicates in §7 there is a "natural Presumption" that what "shall appear evidently impossible and repugnant to one may be presumed the same to another," but does not attempt any further argument on this score. Of course, it is difficult to see how else one might argue for the inconceivability of the object of the calculus and we need not dismiss this objection out of hand. Still, the history of mathematics is replete with instances of supposedly inconceivable objects which later gained mathematical acceptibility as their usefulness became manifest. To take the most obvious examples: irrational, negative, and complex numbers have all been declared inconceivable at one time or another but concerns about their intelligibility soon faded. Berkeley is correct to point out that mathematicians' pronouncements on the nature of fluxions, moments, and infinitesimals conflict with an "official" standard of rigor which holds that the objects of a mathematical demonstration be clearly and distinctly conceived, but this is hardly a devastating criticism. If Berkeley's case against the calculus were confined to simple reports of his inability to frame the appropriate ideas, *The Analyst* would be justly ignored.

The real heart of Berkeley's case against the calculus is his analysis of two Newtonian proofs of fundamental results. These consitute his logical criticism, and they are his attempt to make good on his claim in §8 that when we look carefully into the procedures of the calculus we will find "much Emptiness, Darkness, and Confusion; nay. . . direct Impossibilities and Contradictions." These two Newtonian proofs are analyzed in §§9-19, with the first coming from Book II, §II, Lemma II of the *Principia* and critiqued in §§9-11 of *The Analyst*. In the *Principia*, Newton presents a method for finding the fluxion of a product or "rectangle" of two flowing quantities. Newton treats the product as a the area of a rectangle, whose sides are the flowing quantities A and B; the moments of these flowing quantities are a and b. The proof proceeds by considering the case where the sides both lack one-half their moments and the resulting rectangle has an area of

$$(A - \tfrac{1}{2}a) \times (B - \tfrac{1}{2}b) \qquad [17]$$

By multiplying through, this becomes

$$AB - \tfrac{1}{2}aB - \tfrac{1}{2}bA + \tfrac{1}{4}ab \qquad [18]$$

We then take the rectangle formed after the flowing quantities have been increased by the remaining halves of their moments, viz:

$$(A + \tfrac{1}{2}a) \times (B + \tfrac{1}{2}b) \qquad [19]$$

When exapanded, this becomes

$$AB + \tfrac{1}{2}aB + \tfrac{1}{2}bA + \tfrac{1}{4}ab. \qquad [20]$$

Newton then claims that the moment of the product will be the difference between equations [20] and [18], namely $aB + bA$.

Berkeley's response to this proof is to dismiss it as a sham. He rightly points out that the "direct and true" method of finding the increment of the area is to begin with the product AB and compare it to the product $(A+a) \times (B+b)$. The result is that the augmented rectangle will have an area of

$$AB + aB + bA + ab \qquad [21]$$

so that the momentaneous increment of the area will be $Ab + bA + ab$, a result which differs from Newton's by containing an additional term ab.

Berkeley's analysis reveals a fundamental flaw in the Newtonian procedure. Newton begins his discussion of moments with the declaration that

> These quantities I here consider as variable and in-
> determined, and increasing or decreasing, as it were,
> by a continual motion of flux; and I understand their
> momentary increments or decrements by the name of
> moments; so that the increments may be esteemed as
> added or affirmative moments; and the decrements as
> subtracted or negative ones. (*Principia*, 1: 249)

Here we have a straightforward declaration that the moment of a flowing
quantity is its momentaneous increment; thus, the moment of a product AB must be its momentaneous increment. Now if the moments of
the quantities A and B are a and b, we must conclude that the moment
of the product is the difference between AB and $(A+a) \times (B+b)$, just as
Berkeley contends. Newton's procedure here is utterly mysterious, since
he actually takes the increment of the rectangle $(A - \frac{1}{2}a) \times (B - \frac{1}{2}b)$. Not
only does Newton take the increment of the wrong product here, but his
whole procedure depends upon the confusing supposition that we can
divide these momentaneous increments into parts; but if a moment is an
increment whose magnitude is not considered and which is supposed to
be generated in an instant, it is hard to see how to make sense of this
supposition.

Berkeley develops his argument by insisting that no matter how we
chose to interpret the doctrine of moments, the Newtonian procedure
does require the use of infinitely small quantities and it is in the effort
to avoid apparent commitment to infinitesimals that Newton was led to
give his sophistical "proof" of the product rule:

> The Points or mere Limits of nascent Lines are un-
> doubtedly equal, as having no more Magnitude one
> than another, a Limit as such being no Quantity. If
> by a Momentum you mean more than the very initial
> Limit, it must be either a finite Quantity or an Infini-
> tesimal. But all finite Quantities are expressly excluded
> from the Notion of a Momentum. Therefore the Mo-
> mentum must be an Infinitesimal. And indeed, though
> much Artifice hath been employed to escape or avoid
> the admission of Quantities infinitely small, yet it seems

ineffectual. For ought I see, you can admit no Quantity
as a Medium between a finite Quantity and nothing,
without admitting Infinitesimals. An increment generated in a finite Particle of Time is it self a finite Particle; and cannot therefore be a Momentum. You must
therefore take an Infinitesimal Part of Time wherein to
generate your Momentum. It is said, the Magnitude of
Moments is not considered: And yet these same Moments are supposed to be divided into Parts. This is
not easy to conceive, no more that it is why we should
take Quantities less than A and B in order to obtain
the Increment of AB, of which proceeding it must be
owned the final Cause or Motive is very obvious; but
it is not so obvious or easy to explain a just and legitimate Reason for it, or shew it to be Geometrical.
(*The Analyst*, §11)

There is nothing to contest in this passage, and with it Berkeley has gone
a very long way to establish his central claim for the absence of rigor in
the calculus.[8] Certainly, Newton's mysterious procedure is motivated by
a desire to avoid embarassing questions about infinitesimal magnitudes,
but in setting out a "proof" of this sort Newton has only shown how
unrigorous the calculus really is. We can thus grant that Berkeley is
right on two counts: the procedures of the calculus are *prima facie* not
properly demonstrated and the Newtonian apparatus of fluxions and
moments is indistinguishable from the infinitesimal calculus of Leibniz.
    Having dealt with Newton's *Principia* proof of the product rule,
Berkeley goes on in §12 to consider the rule for finding the fluxion of

---

[8] Commentators have noted the strength of Berkeley's case on this point.
Blay observes: "Or, Newton ne donne aucune justification de sa procédure, si
ce n'est d'une part que $\frac{1}{2}a - (-\frac{1}{2}a) = a$, et d'autre part, qu'elle permet de se
débarraser des termes du deuxième ordre sans les négliger puisqu'ils disparais-
sent d'eux-mêmes dans le calcul newtonien," (Blay 1986, 245-6). Sherry (1987),
although otherwise unsympathetic to Berkeley's case against Newton, grants
that Berkeley's critique of this Newtonian proof succeeds in showing that it is
merely "window dressing." And Breidert (1989, 101) remarks that in this case
"Berkeley hatte offensichtlich ins Schwarze getroffen. Newton hatte nämlich zur
Gewinnung der Productregel auf einen Trick zurückgegriffen, dessen Schwäche
leicht einzusehen ist."

any power, as demonstrated in the treatise *Quadrature of Curves*. Berkeley's interest in this second proof is understandable: the *Quadrature of Curves* is a much more complete statement of the calculus than that which appears in the *Principia* and the method of proof is importantly different. Berkeley insists that the obscurity of the proof in the *Principia* is the result of Newton's attempt to mask his use of infinitesimals, but suspects that Newton must have suffered "some inward Scruple or Consciousness of defect in the foregoing Demonstration." In view of the fundamental importance of the result for the whole calculus, Berkeley suspects that Newton resolved "to demonstrate the same in a manner independent of the foregoing Demonstration."

Berkeley prefaces his objection to the Newtonian proof by stating a lemma which he regards as "so plain as to need no Proof." The lemma reads:

> If with a View to demonstrate any Proposition, a certain Point is supposed, by virtue of which certain other Points are attained; and such supposed Point be it self afterwards destroyed or rejected by a contrary Supposition; in that case, all the other Points attained thereby, and consequent thereupon, must also be destroyed and rejected, so as from thence forward to be no more supposed or applied in the Demonstration. (*The Analyst*, §12)

In essence this lemma asserts that contradictory premises are not to be admitted in a demonstration. As such, it is a completely unexceptionable principle, since to violate it would permit the use of obviously fallacious patterns of argumentation. Berkeley uses this principle as the basis of an objection to the Newtonian demonstration in the Introduction to the *Quadrature of Curves*, where he claims to find Newton employing contradictory assumptions about the quantity o. Newton's proof proceeds as follows: we begin with an equation x^n, with x any fluent and n any power. To compute the fluxion of x^n, we first consider an increment o

of the fluent x. As the fluent acquires its increment, the power $(x + o)^n$ becomes, by binomial expansion, the quantity

$$x^n + nox^{n-1} + \frac{n(n-1)}{2}o^2x^{n-2} + \ldots \qquad [22]$$

We then have two increments, one of the fluent and one of the power:

$$o \quad \text{and} \quad nox^{n-1} + \frac{n(n-1)}{2}o^2x^{n-2} + \ldots \qquad [23]$$

To find the fluxion, we compute their ratio. After dividing through by the common term o that ratio is

$$1 \quad \text{to} \quad nx^{n-1} + \frac{n(n-1)}{2}ox^{n-2} + \ldots \qquad [24]$$

Now we let the quantity o vanish and discard terms containing it to find the ultimate ratio of the evanescent increments. This turns out to be the ratio of 1 to $nx^{(n-1)}$. But the fluxion is just the ultimate ratio of the evanescent increments, and therefore the fluxion of the power x^n must be $nx^{(n-1)}$.

Berkeley contends that the method of reasoning employed in this proof conflicts with his foregoing lemma by making contradictory assumptions concerning o. In the beginning, it is supposed that o is a positive quantity, since the computation of the increments and the comparison of the ratios of these increments both depend upon the supposition that o is greater than zero. But after the ratios of the increments have been simplified by dividing out the common term o, a new and contradictory assumption is made, namely that the quantity o is equal to zero. This new assumption is contrary to the original, and when it is introduced all consequences drawn from the original assumption must be rejected. But, in fact, important consequences are retained – consequences that cannot be derived from the new assumption. Berkeley goes on at great length on this point in §§14 - 16, insisting that this Newtonian method of proof is an entirely sophistical exercise in the "shifting of the Hypothesis." He observes that this fallacious procedure does involve a certain degree of skill, since it is a delicate matter to decide when to shift from the supposition that o is a positive quantity to the new supposition that it is zero. He further points out that the desired result cannot be obtained if this shift is made too early, because either all terms will reduce to zero or a division by zero will be required. But this knack or skill is no substitute for sound and convincing demonstrations, and Berkeley roundly condemns the prevalence of this kind of argument among the proponents of the Newtonian calculus, concluding that it is only among mathematicians that such blatant fallacies are accepted for demonstration.

In §17 Berkeley concludes his attack on the Newton's proof from the *Quadrature of Curves* by declaring that it is essentially the same method employed by the proponents of the differential calculus. He insists that when Newton supposes the increment *o* to be infinitely diminished and then rejects it, he is effectively rejecting an infinitesimal. As Berkeley rightly observes, it requires a "marvellous sharpness of Discernment" to distinguish between an evanescent increment and an infinitesimal difference. Indeed, the role of the mysterious quantity *o* can only be interpreted as that of an infinitesimal quantity, since it is treated as being both positive and less than any assignable magnitude. In the Leibnizian presentation of the calculus, however, there is no attempt to hide any use of infinitesimals, and the inconsistency of the procedure appears all the more clearly. By "making no manner of Scruple" to take *dx* and *dy* as positive magnitudes when we need to divide by them and as zero when we need to discard them, the Continental analysts embrace essentially the same inconsistency as Newton, except only that they are vastly less embarrassed by it.

Berkeley's logical objections to the calculus are both brilliantly argued and devastatingly effective. Florian Cajori characterized them as "so many bombs thrown into the mathematical camp," (Cajori 1919, 57) and his judgement seems correct. At the very least Berkeley can embarrass those defenders of Newton who had proclaimed the superior rigor of the calculus of fluxions over its Continental rival, but in the aftermath of Berkeley's criticisms the Newtonian approach is itself suspect and the justification of the methods of "the Modern Analysis" is a matter of urgency. This is not to say that Berkeley's criticisms are unanswerable. As anyone familiar with the modern theory of limits knows, a rigorous presentation of the calculus is possible within the confines of the theory of sequences, series, and convergence developed in the nineteenth century. Yet these developments came a century after Berkeley's critique of the calculus, so it is takes nothing away from Berkeley's attack on these

methods to say that a response to his objections gained wide acceptance more than a century after the publication of *The Analyst*. In fact,
the aptness of Berkeley's case can be gathered by noting that no figure
in eighteenth-century British mathematics could give a widely-accepted
exposition of the Newtonian theory of prime and ultimate ratios. Far
from it: the efforts of Newton's successors to elucidate the mysteries of
evanescent magnitudes were so notably unsuccessful that they led to a
bitter controversy among British mathematicians in the mid-eighteenth
century. In this dispute avowed Newtonians exchanged vituperative essays intended to set forth the true meaning of the Newtonian doctrine.
The confusion engendered by Newton's pronouncements on the nature
of evanescent magnitudes was so great that John Wright, in his commentary on Newton's *Principia* was moved to declare:

> After all, however, neither [Newton] himself nor any of
> his Commentators, thought much has been advanced
> upon the subject, has obviated this objection. Bishop
> Berkeley's ingenious criticisms in the Analyst remain
> to this day unanswered. He therein facetiously denom-
> inates the results, obtained from the supposition that
> the quantities, before considered finite and real, have
> vanished, the "Ghosts of Departed Quantities;" and it
> must be admitted there is reason as well as wit in the
> appellation. The fact is, Newton himself . . . had no
> knowledge of the *true nature* of his Method of Prime
> and Ultimate Ratios. (Wright [1833] 1972, 2-3)[9]

To find such a commentary surveying the efforts of the century after
the *Analyst* is hardly what we could expect if the Newtonian procedure
was as rigorous and open to favorable interpretation as his defenders
maintained. If anything, the history of the calculus after *The Analyst*
shows that (at least in Britain) the theory was the object of numerous
conflicting interpretations. Moreover, these conflicting interpretations

---

[9] It should be noted that Wright misrepresents Berkeley's famous reference
to "Ghosts of departed Quantities" as applying to the results obtained from the
supposition that finite magnitudes have vanished. But Berkeley characterizes
evanescent magnitudes as ghosts of departed quantities in §35 of *The Analyst*.

were all attempts to find a rigorous presentation of the calculus which could overcome Berkeley's objections, but the lack of agreement on how best to interpret the calculus suggests that he was right to characterize it as a mathematical theory whose principles were obscure and mysterious.

Berkeley follows his critique of the calculus with an attempt to explain how the flawed procedures of the calculus can nevertheless deliver correct results through an elaborate compensation of errors. The foundation of Berkeley's compensation of errors thesis is the tacit assumption that all of the inferences contained in the calculus can be represented within the framework of classical geometry. His strategy is to show how the application of the calculus to the solution of a problem essentially involves a false geometric assumption which is then compensated by a false algebraic assumption when the problem is represented analytically in Cartesian co-ordinates. Unfortunately for Berkeley, the results are not terribly impressive. At best, he can show examples where the new methods are replaceable by the old, but a completely general proof of their eliminability is not forthcoming. But failure in the positive account of how the calculus might nevertheless achieve correct results does not detract from the success of Berkeley's negative case against the new methods of Newton and Leibniz.

4. OUTLINE ANALYSIS

The structure of *The Analyst* follows quite naturally from the rhetorical conventions of the eighteenth-century polemical piece. After a brief introduction (§§1-2), Berkeley launches into an investigation of the "Object, Principles and Demonstrations" of the modern methods (§§3-20). These sections are followed by an extended attempt to explain how the flawed reasoning of the calculus can nevertheless yield true conclusions. Berkeley spends §§21-29 arguing that the procedures of the calculus involve a compensation of errors in which two false suppositions cancel one another and yield and accurate conclusion. Having satisfied himself that the success of the calculus can be explained, Berkeley then anticipates various responses to his critique. He thus devotes §§30-47 to presentations and interpretations of the calculus which might avoid the arguments from §§3-20. In each case, Berkeley contends that the proffered alternative is unsatisfactory. He concludes with a blanket indictment of the modern methods as obscure and unscientific (§§48-50), adding a series of sixty-seven "Queries" which range widely over topics in the philosophy of mathematics and related fields. The following detailed analysis gives a more careful presentation of Berkeley's line of argument.

4.1 INTRODUCTION (§§1-2)

The behavior of an infidel mathematician is decried, who abuses his reputation as a master of reason and argument to mislead unwary persons in matters of religion. Berkeley proposes to turn the tables by taking the privelege of a freethinker and conducting a dispassionate analysis of the "Object, Principles, and Method of Demonstration admitted by the Mathematicians of the present Age." He admits (§2) that geometry, when guided by proper principles, exemplifies sound reasoning, but suggests that the "Geometrical Analysts" who employ the calculus fail to live up to this standard.

4.2 THE OBJECT OF THE CALCULUS (§§3-8)

The Newtonian calculus of fluxions is examined to see whether its object is readily conceivable (§§3-4). Berkeley observes that fluxions are defined as the velocities with which geometric magnitudes are produced, but such velocities are defined as the first ratios of nascent increments. Further, the calculus of fluxions introduces the doctrine of moments, understood as "the momentaneous Increments or Decrements of undetermined flowing Quantities." But all of these notions are obscure and unscientific: moments are neither finite particles, nor nothing, nor infinitesimals; fluxions are velocities with which nascent increments are produced, but these increments are inconceivable. Moreover, there are fluxions of fluxions, which appear to be even more mysterious. The Leibnizian differential calculus is then considered (§§5-6) and found to be as incomprehensible as the doctrine of fluxions. The notion of infinitesimal magnitudes is rejected as obscure (§5), to say nothing of the further obscurities involved in the doctrine of higher-order differentials (§6). Berkeley grants that the proponents claim to find no difficulty in conceiving the object of their science, but suggests that they have only constructed a notation for fluxions and infinitesimals without conceiving any objects denoted by the various signs and symbols (§§7-8).

4.3 THE PRINCIPLES AND DEMONSTRATIONS OF THE MODERN ANALYSIS (§§9-20)

Berkeley criticizes the Newtonian proof of the rule for computing the fluxion of a product of two fluents as it is presented in the *Principia* (§§9-11). He observes that the Newtonian proof depends upon an obscure supposition which yields a false result and is equivalent to rejecting an infinitesimal quantity. He then takes up the more general case for computing the fluxion of any power, concentrating on a proof drawn from the *Quadrature of Curves* (§§12-17). He lays down a lemma which, in effect, bars the use of inconsistent premises in a demonstration (§12), then presents Newton's reasoning and argues that Newton offends against the lemma by assuming that an increment is first positive and then zero (§§13-16). Berkeley declares that Newton's suppositions are essentially equivalent to those of the differential calculus (§17), observing that the Leibnizians "make no manner of Scruple" to introduce and dismiss infinitesimal magnitudes at pleasure (§18). He ends this part of the argument by noting that the principles of the calculus cannot be vindicated by appeal to the truth of its conclusions (§19), and insists that the "Geometrical Analyst" is to be considered as a logician (§20).

4.4 THE COMPENSATION OF ERRORS THESIS (§§21-29)

Berkeley shifts to an explanation of how the flawed reasoning of the calculus can yield true results. He takes an example from the *calculus differentialis* and argues that the method of computing the subtangent to a parabola involves two compensating errors which balance (§§21-23). A second example computes the value for the subtangent by appearing to rely upon one infinitesimal, but again introducting compensating errors (§§24-25). Berkeley then considers an application of the theory of evanescent increments to find the ordinate of a curve, arguing again that the rejected evanescent quantity is actually cancelled by another with the same magnitude and an opposite sign (§§26-27). This case is then generalized to cover any power, and Berkeley argues that in each case there is a balancing of geometrical and algebraic quantities which invariably yield the correct result (§§28-29).

4.5 ALTERNATIVE INTERPRETATIONS OF THE CALCULUS REJECTED (§§30-47)

The critique of the doctrine of fluxions is extended by considering and rejecting various possible interpretations of it. In §§30-31 Berkeley insists that the doctrine of evanescent increments and the theory of differences are unacceptable because we cannot conceive a ratio between two quantities that have vanished, nor can we imagine a velocity abstracted from time and space in order to compute the ratio between two velocities. He then supposes that some will claim that the practical application of fluxions does not require one to comprehend these difficult suppositions (§32), but rejects this as abandoning the ideal of science and demonstration (§33). In §34 he considers Newton's claim that fluxions can be expressed by the proportions of finite lines; but this depends upon the absurd supposition that a point can be considered as a triangle. A purely formalistic interpretation of the calculus is then (§35) proposed, but rejected because it requires the same fallacious reasoning as that critiqued in §15. Higher-order fluxions are then interpreted as a sequence of velocities with which successive increments to a finite right line are generated (§36), but this simply introduces a sequence of algebraic expressions to which no distinct ideas can be found to correspond (§37).

Berkeley takes up the suggestion that fluxions of all orders might be interpreted as the velocities with which infinitesimal increments are generated, but rejects this as inconceivable and inconsistent with Newton's pronouncements on the nature of fluxions (§38). Another alternative would treat all orders of fluxions as the velocities of nascent increments (§§39-41), but this reduces to the same doctrine critiqued in §37. All attempts to interpret fluxions as velocities are then rejected because they require us to frame an abstract idea of *instantaneous* velocity, exclusive of time, space, and duration (§§42-44). The various accounts of fluxions are then rejected (§§45-47) because they provide no insight into the true nature of fluxions themselves, but instead introduce lines, areas, or algebraic symbols to stand for the incomprehensible fluxions.

4.6 CONCLUSION AND QUERIES (§§48-50, Queries 1-67)

The case against the calculus is summarized: analysts are accused of employing "the most incomprehensible Metaphysics," (§48) and proceeding to reason inconsistently about inconceivable entities (§49). The result is that the modern methods fail to meet both metaphysical and logical criteria of rigor, and skill in the calculus does not merit a reputation as a master of reason and argument. Berkeley then (§50) links *The Analyst* to his earlier remarks on mathematics in his 1710 *Treatise Concerning the Principles of Human Knowledge* and states as the occasion for *The Analyst* his "being called upon to make good my Suggestions" in his earlier work, as well as a desire to prevent the infidel mathematician from doing damage to the cause of revealed religion. The Queries which follow are intended to give a fuller exposition of the Berkeleyan philosophy of mathematics. These are clearly modeled on the famous list of "Queries" at the end of Newton's *Opticks* and serve to round out Berkeley's critique by linking his thoughts on the calculus to his conception of geometry, arithmetic, and algebra.

5. RESPONSES TO *THE ANALYST*

Rebuttals to *The Analyst* appeared almost immediately after its publication. Two of these drew responses from Berkeley: James Jurin's *Geometry no Friend to Infidelity* and John Walton's *Vindication of Sir Isaac Newton's Principles of Fluxions*. Berkeley's 1735 *Defence of Free-Thinking in Mathematics* (*Works* 4: 109-141) was principally directed at Jurin, but added an appendix replying to Walton. This Berkeleyan tract drew two answers: Jurin's vituperative piece *The Minute Mathematician; or, the Free-thinker no Just-thinker* and Walton's *Chatecism of the Author of the "Minute Philosopher" fully answer'd.* Berkeley took no notice of Jurin's second effort, but he concluded his part in this battle of the books with his last publication on the calculus: *Reasons for not replying to Mr. Walton's Full Answer* (*Works* 4: 147-156). Predictably, Walton replied with a second edition of his *Full Answer* which contained an appendix replying to Berkeley's *Reasons for not replying.* The final salvo in this battle of the books came in 1736 from the pen of John Hanna, who critiqued Walton's efforts in defense of Newton with *Some Remarks on Mr. Walton's appendix, which he Wrote in Reply to the Author of the "Minute Philosopher".*

In addition to these works there were numerous replies to *The Analyst* which drew no response from Berkeley. In fact, it is difficult to find a text on the calculus of fluxions from the period 1734-1750 which does not contain at least a veiled reference to *The Analyst.*[10] One of the more important of these replies is Benjamin Robins's 1735 *Discourse Concerning the Nature and Certainty of Sir Isaac Newton's Method of Fluxions, and of Prime and Ultimate Ratios.* This was critiqued by Jurin in an essay in a journal called *The Present State of the Republick of Letters* and the result was a controversy between Jurin and Robins which lasted for several years and occupied hundreds of pages in that journal. The Rev. Thomas Bayes attacked Berkeley in 1736 with an anonymous *Introduction to the Doctrine of Fluxions, and Defence of the Mathematicians against the Objections of the Author of the "Analyst",* while James Smith undertook an entirely different defence of the calculus with his 1737 *New Treatise of Fluxions.* John Colson prepared a translation of Newton's *Method of Fluxions and Infinite Series* in 1736 whose preface and commentary contained an extended attack on Berkeley, and Colin Maclaurin's *Treatise of Fluxions* (1742) attempted to answer Berkeley by setting the calculus of fluxions on the same footing as the classical method of exhaustion. Francis Blake published an anonymous *Explanation of Fluxions* which went through several editions, while Roger Paman presented several papers to the Royal Society which were published in 1745 as *The Harmony of the Ancient and Modern Geometry Asserted.* The volume of responses and the vastly different approaches undertaken

---

[10] For example, see the textbooks Hodgson (1736) and Emerson (1743). The literature from this period is summarized in Cajori (1919) and Guicciardini (1990).

by various authors in the defense of the Newtonian doctrine convinced Berkeley that his criticisms of the calculus had been essentially correct. In the *Siris* (1744) Berkeley observes that

> Our judgement in these matters is not to be overborne by a presumed evidence of mathematical notions and reasonsing, since it is plain the mathematicians of this age embrace obscure notions and uncertain opinions, and are puzzled about them, contradicting each other and disputing like other men: witness their doctrine of fluxions, about which, within these ten years, I have seen published about twenty tracts and dissertaions, whose authors being utterly at variance, and inconsistent with each other, instruct bystanders what to think of their pretensions to evidence. (*Works* 5: 127)

In fact, some of the responses (notably those of Robins, Paman, and Maclaurin) were ingenious and mathematically sound proposals for rigorizing the calculus, but Berkeley seems to have thought little of them. Nevertheless, his critique of the calculus was an important event in the history of British mathematics and philosophy. As George Gibson concluded a century ago, "Berkeley did great service to sound reasoning in mathematics by the publication of *The Analyst*. Were it for nothing else than the *Discourse* and *Dissertaion* of Robins and the *Fluxions* of Maclaurin, Berkeley's name should be had in reverence of mathematicians," (Gibson 1891, 31-2).

Aside from its purely mathematical interest, Berkeley's *Analyst* is of obvious importance to historians of philosophy. Whatever the final status of his critique of the calculus, anyone who would understand Berkeley's conception of mathematics and his criteria for mathematical rigor must follow the argument of *The Analyst*. Thus, to the extent that an understanding of Berkeley's philosophy of mathematics is essential for an appreciation of his philosophy as a whole, *The Analyst* is a text which must be read and understood by Berkeley scholars. Furthermore, although *The Analyst* makes no mention of immaterialism, Berkeley's procedure is rooted in epistemological doctrines familiar from his other works, and we should not treat it as an isolated foray into mathematics

which lacks any deep connection with other parts of the Berkeleyan
enterprise.

6. A NOTE ON THE TEXT

The present edition is based upon the first edition (London: J. Tonson, 1734). Another edition was published in Dublin by S. Fuller and
J. Leathly later the same year, reproducing the London edition with
only changes in spelling, capitalization, and punctuation. A posthumous second London edition appeared in 1754 (again published by J.
Tonson) and is essentially a re-issue of the 1734 London edition. The
1734 Dublin edition is much rarer than its London counterpart and contains significantly more typographical errors, thus dictating the use of
the 1734 London edition as copy-text. The present edition corrects the
errata of the original and follows its punctuation and capitalization, but
where Berkeley used roman numerals in section numbering I have used
the character '§' with arabic numerals. I have replaced the long 's' with
the short 's' but have otherwise left the text unaltered. The mathematical notation has been slightly revised for the sake of readability: where
Berkeley has '$\overline{x+o}|^n$' I use the modern '$(x+o)^n$' and I have replaced
such expressions as 'nn' with the exponential notation 'n^2'. I have re-
drawn Berkeley's figures to conform to modern conventions concerning
the orientation of axes, labeling each figure and placing references to
the figures in square brackets in the text. Berkeley's footnotes are reproduced, with the asterisk and (where necessary) dagger character as
reference marks. In some cases I have added explanatory material to the
original notes, placing this in square brackets at the end of the original
note. My own notes are numbered consecutively throughout the text.

7. BIBLIOGRAPHY

I have tried to be reasonably comprehensive in compiling this bibliography, but make no claim to completeness. The rationale for selection is to
include all of the primary literature immediately relevant to Berkeley's

Analyst and responses to it. The secondary literature deals principially with Berkeley's mathematical writings, but includes some general histories and studies of particular figures which are part of the mathematical background to Berkeley. Since bibliographies in the history of mathematics tend to ignore Berkeley's *Analyst* while bibliographies of Berkeley often leave out mathematical work, I have tried to bridge the gap by including all and only those works necessary for an understanding of *The Analyst*.

Andersen, Kirsti. 1985. Cavalieri's Method of Indivisibles. *Archive for History of the Exact Sciences* 24: 292-367.

Apollonius of Perga. [1939] 1952. *Conics*. Trans. R. Catesby Taliaferro. Reprinted in *Great Books of the Western World* 11: 593-804. Chicago: Encyclopædia Brittanica.

Baron, Margaret E. 1969. *The Origins of the Infinitesimal Calculus*. Oxford: Pergamon Press.

Barrow, Isaac. 1860. *The Mathematical Works of Isaac Barrow, D. D.* Ed. W. Whewell. Two vols. bound as one. Cambridge: Cambridge University Press.

Baum, Robert J. 1969. *George Berkeley's Philosophy of Mathematics*. Ph.D. diss., Ohio State University. Ann Arbor, MI: University Microfilms.

Baum, Robert J. 1972. Instrumentalist and Formalist Elements in Berkeley's Philosophy of Mathematics. *Studies in History and Philosophy of Science* 3: 119-34.

[Baxter, Andrew]. 1733. *An Enquiry into the Nature of the Human Soul; wherein the Immateriality of the Soul Is evinced from the Principles of Reason and Philosophy*. London: G. Strahan.

[Bayes, Rev. Thomas]. 1736. *An Introduction to the Doctrine of Fluxions, and Defence of the Mathematicians Against the Objections of the Author of the "Analyst," so far as they are Designed to Affect their General Methods of Reasoning*. London: J. Noon.

Berkeley, George. 1948-1957. *The Works of George Berkeley, Bishop of Cloyne*. Ed. A.A. Luce and T.E. Jessop. Edinburgh and London: Nelson.

Berlioz, Dominique. 1988. Berkeley et la polémique du calcul infinitésimal. In *Entre forme et histoire: la formation de la notion de développement à l'âge classique*, ed. Olivier Bloch, Bernard Balan, and Paulette Carrive, 71-85. Paris: Meridiens Klincksieck.

[Blake, Francis]. 1741. *An Explanation of Fluxions, in a Short Essay on the Theory*. London: W. Innys.

Blay, Michel. 1986. Deux Moments de la critique du calcul infinitésimal: Michel Rolle et George Berkeley. *Revue d'histoire des sciences* 39: 223-53.

Bos, H.J.M. 1974. Differentials, higher-order differentials and the derivative in the Leibnizian calculus. *Archive for History of the Exact Sciences* 14: 1-90.

Boyer, Carl B. [1949] 1959. *The History of the Calculus and its Conceptual Development*. New York: Dover.

Boyer, Carl B. [1968] 1989. *A History of Mathematics*. 2d ed. rev. Eta Merzbach. New York: John Wiley & Sons.

Breidert, Wolfgang. 1986. Berkeley's Kritik an der Infinitesimalrechnung. In Heinekamp (1986), 185-90.

Breidert, Wolfgang. 1989. *George Berkeley, 1685-1753* Vita Mathematica, 4. Basel, Boston, and Berlin: Birkhäuser.

Cajori, Florian. 1917. Discussion of Fluxions from Berkeley to Woodhouse. *American Mathematical Monthly* 24: 145-54.

Cajori, Florian. 1919. *A History of the Conceptions of Limits and Fluxions in Great Britain from Newton to Woodhouse*. Chicago and London: Open Court.

Cajori, Florian. 1925. Indivisibles and "Ghosts of Departed Quantities" in the History of Mathematics. *Scientia* 37: 301-06.

Cantor, Geoffrey. 1984. Berkeley's *Analyst* Revisited. *Isis* 75: 668-83.

Claussen, Friedrich. 1889. *Kritische Darstellung der Lehren Berkeleys über Mathematik und Naturwissenschaften*. Halle a.S.: Erhardt Karras.

[Collins, Anthony.] 1713. A discourse of free-thinking, occasion'd by the rise and growth of a sect call'd free-thinkers. London.

Colson, John. 1736. See (Newton 1736).

De Morgan, Augustus. 1852. On the Early History of Infinitesimals in England. *London, Edinburgh and Dublin Philosophical Magazine and Journal of Science*. 4: 321-30.

Devaux, Philippe. 1953. Berkeley et les mathématiques. *Revue internationale de philosophie* 7: 101-33.

Edwards, C.H. 1979. *The Historical Development of the Calculus*. New York and Berlin: Springer Verlag.

Emerson, William. 1743. *The Doctrine of Fluxions: not only Explaining the Elements thereof, but also its Application and Use in the Several Parts of Mathematics and Natural Philosophy*. London: Bettenham.

Euclid. [1925] 1956. *The Thirteen Books of Euclid's "Elements" Translated from the Text of Heiberg*. Ed. and trans. T.L. Heath. 3 vols. New York: Dover.

Evans, W. D. 1914. Berkeley and Newton. *Mathematical Gazette* 7: 418-21.

Fauvel, John and Jeremy Gray, eds. 1987. *The History of Mathematics: A Reader*. Basingstoke, Hampshire: MacMillan.

Gibson, George A. 1898. *Vorlesungen über Geschichte der Mathematik von Moritz Cantor. Dritter Band, Dritte Abteilung*. A Review: with special reference to the *Analyst* controversy. *Proceedings of the Edinburugh Mathematical Society* 17: 9-32.

Gibson, George A. 1899. Berkeley's *Analyst* and its critics: an episode in the development of the doctrine of limits. *Bibliotheca Mathematica* 13: 65-70.

Giorello, Giulio. 1985. *Lo Spettro e il Libertino: Teologia, Matematica, Libero Pensiero*. Milan: Mondadori.

Giusti, Enrico. 1980. *Bonaventura Cavalieri and the Theory of Indivisibles*. Milan: Edizioni Cremonese.

Grabiner, Judith. 1981. *The origins of Cauchy's rigorous calculus*. Cambridge, MA: MIT Press.

Grattan-Guinness, I. 1969. Berkeley's Criticism of the Calculus as a Study in the Theory of Limits. *Janus* 56: 215-227.

Grattan-Guinness, I., ed. 1980. *From the Calculus to Set Theory, 1630-1910: An Intoductory History*. London: Duckworth.

Guicciardini, Niccolò. 1990. *The development of the Newtonian calculus in Britain: 1700-1800*. Cambridge: Cambridge University Press.

Hall, A. R. 1980. *Philosophers at War: The Quarrel Between Newton and Leibniz*. Cambridge: Cambridge University Press.

Hanna, John. 1736. *Some Remarks on Mr. Walton's Appendix, which he Wrote in Reply to the Author of the "Minute Philosopher": concerning Motion and Velocity*. Dublin: Fuller.

Heinekamp, Albert, ed. 1986. *300 Jahre "Nova methodus" von G.W. Leibniz (1684-1984)*. Symposium of the Leibniz-Gesellschaft, *Studia Leibnitiana* Sonderheft no. 14. Stuttgart: Franz Steiner Verlag.

Hodgson, James. 1736. *The Doctrine of Fluxions, Founded on Sir Isaac Newton's Method, Published by Himself in his Tract upon the Quadrature of Curves*. London: T. Wood.

Jesseph, Douglas M. 1989. Philosophical Theory and Mathematical Practice in the Seventeenth Century. *Studies in History and Philosophy of Science* 20: 215-244.

Jesseph, Douglas M. 1990. Berkeley's Philosophy of Geometry. *Archiv für Geschichte der Philosophie* 72: 301-322.

Johnston, G. A. 1916. The Influence of Mathematical Conceptions on Berkeley's Philosophy. *Mind* 25: 177-92.

Johnston, G. A. 1918. Berkeley's Logic of Mathematics. *The Monist* 28: 25-45.

Johnston, G. A. 1923. *The Development of Berkeley's Philosophy* London: Macmillan.

[Jurin, James]. 1734. *Geometry no Friend to Infidelity; or, a Defence of Sir Isaac Newton and the British Mathematicians, in a Letter to the Author of the "Analyst": Wherein it is Examined how far the Conduct of such Divines as Intermix the Interest of Religion with their Private Disputes and Passions, and Allow Neither Learning nor Reason to those they differ from, is of Honour or Service to Christianity, or Agreeable to the Example of our Blessed Savior and his Apostles*. London: T. Cooper.

[Jurin, James]. 1735. *The Minute Mathematician; or, the Free-Thinker no Just-Thinker. Set forth in a Second Letter to the Author of the "Analyst"; Containing a Defence of Sir Isaac Newton and the British Mathematicians, against a late Pamphlet, entituled "A Defence of Free-Thinking in Mathematicks"*. London: T. Cooper.

Keynes, Geoffrey, Kt. 1976. *A Bibliography of George Berkeley, Bishop of Cloyne: His works and his Critics in the Eighteenth Century*. Oxford: Clarendon Press.

Kitcher, Philip. 1973. Fluxions, limits, and infinite littlenesse: a Study of Newton's presentation of the calculus. *Isis* 44: 33-49.

Kitcher, Philip. 1984. *The Nature of Mathematical Knowledge*. Oxford and New York: Oxford University Press.

Leibniz, G.W. [1848-1853] 1962. *G.W. Leibniz Mathematische Schriften*. 7 vols. ed. C.I. Gerhardt. Hildesheim: Georg Olms.

L'Hôpital, G.F.A. 1696. *Analyse des infiniment petits pour l'intelligence des lignes courbes*. Paris: Imprimiere Royale.

Maclaurin, Colin. 1742. *A Treatise of Fluxions, in Two Books*. 2 vols. Edinburgh: Ruddimans.

Meyer, Eugen. 1894. *Humes und Berkeleys Philosophie der Mathematik, vergleichend und kritisch dargestellt*. Abhandlungen zur Philosophie und ihrer Geschichte, no. 3. Halle a.S: Max Niemeyer.

Newton, Sir Isaac. 1736. *The Method of Fluxions and Infinite Series: with its Application to the Geometry of Curve-Lines. By the Inventor Sir Isaac Newton, Kt. Late President of the Royal Society. Translated from the Author's Latin Original not yet made publick, To which is subjoin'd, A Perpetual Comment upon the whole Work, Consisting of Annotations, Illustrations, and supplements, In order to make this Treatise a Compleat Institution for the use of Learners*. Ed. and trans. John Colson. London: Woodfall.

Newton, Sir Isaac. [1729] 1934. *Sir Isaac Newton's Mathematical Principles of Natural Philosophy and his System of the World*. 2 vols. Trans. Andrew Motte rev. and ed. Florian Cajori. Berkeley, CA: University of California Press.

Newton, Sir Isaac. 1967-1981. *The Mathematical Papers of Isaac Newton*. Ed. D.T. Whiteside and M.A. Hoskins. 8 vols. Cambridge: Cambridge University Press.

Paman, Roger. 1745. *The Harmony of the Ancient and Modern Geometry Asserted: In Answer to the Call of the Author of the "Analyst" upon the Celebrated Mathematicians of the Present Age, to clear up what he Stiles, their Obscure Analytics*. London: J. Nourse.

Pycior, Helena. 1987. Mathematics and Philosophy: Wallis, Hobbes, Barrow, and Berkeley. *Journal of the History of Ideas* 48: 265-86.

Rigaud, Stephen J., ed. [1841] 1965. *Correspondence of scientific men of the seventeenth century; Including letters of Barrow, Flamsteed, Wallis and Newton, printed from originals in the collection of the Right Honourable Earl of Macclesfield*. Two vols. Hildesheim: Georg Olms.

Robins, Benjamin. 1735. *A Discourse Concerning the Nature and Certainty of Sir Isaac Newton's Methods of Fluxions, and of Prime and Ultimate Ratios*. London: W. Innys.

Robins, Benjamin. 1761. *Mathematical Tracts of the late Benjamin Robins, Esq. Fellow of the Royal Society, and Engineer General to the Honourable the East India Company*. Ed. James Wilson. 2 vol. London: J. Nourse.

Robles, José A. 1980. Percepcion y Infinitesimales en Berkeley. *Dianoia* 26: 151-77.

Robles, José A. 1981. Percepcion y Infinitesimales en Berkeley II. *Dianoia* 27: 166-85.

Robles, José A. 1984. Berkeley y su Critica a los Fundamentos del Cálculo. *Rivista Latinoamericana de Filosofia*. 10: 141-50.

Sherry, David. 1987. The Wake of Berkeley's Analyst: Rigor Mathematicæ? *Studies in History and Philosophy of Science* 18: 455-480.

Smith, G. C. 1980. Thomas Bayes and Fluxions. *Historia Mathematica* 7: 379-388.

Smith, James. 1769. *A New Treatise of Fluxions*. London: printed for the author.

Stammler, Gerhardt. 1921. *G. Berkeleys Philosophie der Mathematik*. *Kant-Studien*, Ergänzungsheft no. 55. Berlin: Reuther & Reichard.

Stock, Joseph. [1776] 1989. *An Account of the Life of George Berkeley, D. D. Late Bishop of Cloyne in Ireland. With Notes, Containing Strictures Upon his Works*. Reprint in David Berman, ed. 1989. *George Berkeley: Eighteenth Century Responses*. 2 vols. New York: Garland, 1: 5-85.

Strong, Edward W. 1957. Mathematical Reasoning and its Objects. In *George Berkeley: Lectures delivered before the Philosophical Union of the University of California in honor of the two hundredth anniversary of the death of George Berkeley, Bishop of Cloyne*. Ed. Steven C. Pepper, Karl Aschenbrenner and Benson Mates, 65-88. University of California Publications in Philosophy no. 29. Berkeley and Los Angeles: University of California Press.

Struik, D. J., ed. [1969] 1986. *A Source Book in Mathematics, 1200-1800*. Princeton, NJ: Princeton University Press.

Taylor, Brook. 1715. *Methodus Incrementorum Directa & Inversa*. London: W. Innys.

Tindall, Matthew. 1730. *Christianity as old as the creation: or, The gospel, a republication of the religion of nature*. London.

Toland, John. *Christianity not mysterious: or, A treatise shewing that there is nothing in the gospel contrary to reason, nor above it: and that no Christian doctrine can be properly call'd a mystery*. 2nd ed. London: Buckley.

Vermeulen, Ben. 1985. Berkeley and Nieuwentijt on Infinitesimals. *Berkeley Newsletter* 8: 1-7.

Wallis, John. 1693-99. *Johannis Wallis S. T. D. . . Opera Mathematica*. 3. vols. Oxford: At the Sheldonian Theater.

Walton, John. 1735a. *A vindication of Sir Isaac Newton's Principles of Fluxions, against the Objections contained in the "Analyst"*. Dublin: J. Powell.

Walton, John. 1735b. *The Chatecism of the Author of the "Minute Philosopher" Fully Answer'd*. Dublin: S. Powell.

Walton, John. 1735c. *The Chatecism of the Author of the "Minute Philosopher" Fully Answer'd; With an Appendix in Answer to the "Reasons for not Replying to Mr. Walton's Full Answer"*. Dublin: S. Powell.

Wisdom, John O. 1939. The *Analyst* Controversy: Berkeley's Influence on the Development of Mathematics. *Hermathena* 29: 3-29.

Wisdom, John O. 1941. The Compensation of Errors in the Method of Fluxions. *Hermathena* 57: 49-81.

Wisdom, John O. 1942. The *Analyst* Controversy: Berkeley as Mathematician. *Hermathena* 59: 111-28.

Wisdom, John O. 1953. Berkeley's Criticism of the Infinitesimal. *British Journal for the Philosophy of Science* 4: 22-25.

Wright, J. M. F. [1833] 1972. *A Commentary on Newton's "Principia" with a Supplementary Volume Designed for the Use of Students at the Universities*. The Sources of Science, ed. Harry Woolf, no. 124. New York and London: Johnson Reprint Corp.

THE
ANALYST;
OR, A
DISCOURSE

Addreſſed to an

Infidel MATHEMATICIAN.

WHEREIN

It is examined whether the Object, Principles, and Inferences of the modern Analy-
ſis are more diſtinctly conceived, or more
evidently deduced, than Religious Myſteries
and Points of Faith.

By the AUTHOR of *The Minute Philoſopher.*

G. Berkeley

*Firſt caſt out the beam out of thine own Eye; and then
ſhalt thou ſee clearly to caſt out the mote out of thy bro-
ther's eye.* S. Matt. c. vii. v. 5.

LONDON:
Printed for J. TONSON in the *Strand.* 1734.

THE
CONTENTS

The Analyst

§1. Though I am a Stranger to your Person, yet I am not, Sir, a Stranger to the Reputation you have acquired, in that branch of Learning which hath been your peculiar Study; nor to the Authority that you therefore assume in things foreign to your Profession, nor to the Abuse that you, and too many more of the like Character, are known to make of such undue Authority, to the misleading of unwary Persons in matters of the highest Concernment, and whereof your mathematical Knowledge can by no means qualify you to be a competent Judge. Equity indeed and good Sense would incline one to disregard the Judgement of Men, in Points which they have not considered or examined. But several who make the loudest Claim to those Qualities, do, nevertheless, the very thing they would seem to despise, clothing themselves in the Livery of other Mens Opinions, and putting on a general deference for the Judgement of you, Gentlemen, who are presumed to be of all Men the greatest Masters of Reason, to be most conversant about distinct Ideas, and never to take things upon trust, but always clearly to see your way, as Men whose constant Employment is the deducing Truth by the justest inference from the most evident Principles. With this bias on their Minds, they submit to your Decisions where you have no right to decide. And that this is one short way of making Infidels I am credibly informed.

§2. Whereas then it is supposed, that you apprehend more distinctly, consider more closely, infer more justly, conclude more accurately than other Men, and that you are therefore less religious because more judicious, I shall claim the privilege of a Free-Thinker; and take the Liberty to inquire into the Object, Principles, and Method of Demonstration

admitted by the Mathematicians of the present Age, with the same free-
dom that you presume to treat the Principles and Mysteries of Religion;
to the end, that all Men may see what right you have to lead, or what
Encouragement others have to follow you. It hath been an old remark
that Geometry is an excellent Logic. And it must be owned, that when
the Definitions are clear; when the Postulata cannot be refused, nor the
Axioms denied; when from the distinct Contemplation and Comparison
of Figures, their Properties are derived, by a perpetual well-connected
chain of Consequences, the Objects being still kept in view, and the attention ever fixed upon them; there is acquired an habit of Reasoning,
close and exact and methodical: which habit strengthens and sharpens
the Mind, and being transferred to other Subjects, is of general use in
the inquiry after Truth. But how far this is the case of our Geometrical
Analysts, it may be worth while to consider.

§3. The Method of Fluxions is the general Key, by help whereof
the modern Mathematicians unlock the secrets of Geometry, and consequently of Nature. And as it is that which hath enabled them so
remarkably to outgo the Ancients in discovering Theorems and solving
Problems, the exercise and application thereof is become the main, if
not the sole, employment of all those who in this Age pass for profound
Geometers. But whether this Method be clear or obscure, consistent or
repugnant, demonstrative or precarious, as I shall inquire with the ut-
most impartiality, so I submit my inquiry to your own Judgement, and
that of every candid Reader. Lines are supposed to be generated* by
the motion of Points, Plains by the motion of Lines, and Solids by the
motion of Plains. And whereas Quantities generated in equal times are

---

* Intro. ad Quadraturam Curvaram. [Berkeley's reference here is to the
"Introduction" to Newton's treatise *On the Quadrature of Curves*; the relevant
passage is quoted above in the "Editor's Introduction," Section 1.3. Berkeley's
paraphrase is quite close to the text in (*Papers*, 8:123-129).]

greater or lesser, according to the greater or lesser Velocity, wherewith they increase and are generated, a Method hath been found to determine Quantities from the Velocities of their generating Motions. And such Velocities are called Fluxions: and the Quantities generated are called flowing Quantities. These Fluxions are said to be nearly as the Increments of the flowing Quantities, generated in the least equal Particles of time; and to be accurately in the first Proportion of the nascent, or in the last of the evanescent, Increments. Sometimes, instead of Velocities, the momentaneous Increments or Decrements of undetermined flowing Quantities are considered, under the Appellation of Moments.

§4. By Moments we are not to understand finite Particles. These are said not to be Moments, but Quantities generated from Moments, which last are only the nascent Principles of finite Quantities. It is said, that the minutest Errors are not to be neglected in Mathematics: that the Fluxions are Celerities, not proportional to the finite Increments though ever so small; but only to the Moments or nascent Increments, whereof the Proportion alone, and not the Magnitude, is considered.[1] And of the aforesaid Fluxions there be other Fluxions, which Fluxions of Fluxions are called second Fluxions. And the Fluxions of these second Fluxions are called third Fluxions: and so on, fourth, fifth, sixth, &c. ad infinitum. Now as our Sense is strained and puzzled with the perception of Objects extremely minute, even so the Imagination, which

---

[1] These three sentences paraphrase Newton's *Principia* and *Quadrature of Curves*. In Book II, Lemma II of the *Principia,* Newton writes: "But take care not to look upon finite particles as such. finite particles are not moments, but the very quantities generated by the moments. We are to conceive them as the just nascent principles of finite magnitudes. Nor do we in this Lemma regard the magnitude of the moments, but their first proportion, as nascent," (*Principia,* 1: 249). In the "Introduction" to his *Quadrature of Curves,* Newton insists that "The most minute errors are not in mathematical matters to be scorned." This passage and its Newtonian context will become more important in §§9 and 34.

Faculty derives from Sense, is very much strained and puzzled to frame clear Ideas of the least Particles of time, or the least Increments generated therein: and much more so to comprehend the Moments, or those Increments of the flowing Quantities in *statu nascenti*, in their very first origin or beginning to exist, before they become finite Particles. And it seems still more difficult, to conceive the abstracted Velocities of such nascent imperfect Entities. But the Velocities of the Velocities, the second, third, fourth and fifth Velocities, *&c.* exceed, if I mistake not, all Humane Understanding. The further the Mind analyseth and pursueth these fugitive Ideas, the more it is lost and bewildered; the Objects, at first fleeting and minute, soon vanishing out of sight. Certainly in any Sense a second or third Fluxion seems an obscure Mystery. The incipient Celerity of an incipient Celerity, the nascent Augment of a nascent Augment, *i.e.* of a thing which hath no Magnitude: Take it in which light you please, the clear Conception of it will, if I mistake not, be found impossible, whether it be so or no I appeal to the trial of every thinking Reader. And if a second Fluxion be inconceivable, what are we to think of third, fourth, fifth Fluxions, and so onward without end?

§5. The foreign Mathematicians are supposed by some, even of our own, to proceed in a manner, less accurate perhaps and geometrical, yet more intelligible. Instead of flowing Quantities and their Fluxions, they consider the variable finite Quantities as increasing or diminishing by the continual Addition or Subduction of infinitely small Quantities. Instead of the Velocities wherewith Increments are generated, they consider the Increments or Decrements themselves, which they call Differences, and which are supposed to be infinitely small. The Difference of a Line is an infinitely little Line; of a Plain an infinitely little Plain. They suppose finite Quantities to consist of Parts infinitely little, and Curves to be Polygones, whereof the Sides are infinitely little, which by the Angles

they make one with another determine the Curvity of the Line.[2] Now
to conceive a Quantity infinitely small, that is, infinitely less than any
sensible or imaginable Quantity, or than any the least finite Magnitude,
is, I confess, above my Capacity. But to conceive a Part of such infinitely
small Quantity, that shall be still infinitely less than it, and consequently
though multiply'd infinitely shall never equal the minutest finite Quantity, is, I suspect, an infinite Difficulty to any Man whatsoever; and will
be allowed such by those who candidly say what they think; provided
they really think and reflect, and do not take things upon trust.

§6. And yet in the *calculus differentialis,* which Method serves to all
the same Intents and Ends with that of Fluxions, our modern Analysts
are not content to consider only the Differences of finite Quantities: they
also consider the Differences of those Differences, and the Differences of
the Differences of the first Differences. And so on *ad infinitum.* That is,
they consider Quantities infinitely less than the least discernible Quantity; and others infinitely less than those infinitely small ones; and still
others infinitely less than the preceeding Infinitesimals, and so on without end or limit. Insomuch that we are to admit an infinite succession
of Infinitesimals, each infinitely less than the foregoing, and infinitely
greater than the following. As there are first, second, third, fourth, fifth,
&c. Fluxions, so there are Differences, first, second, third, fourth, *&c.,*
in an infinite Progression towards nothing, which you still approach and

---

[2] Berkeley is here paraphrasing the Marquis de L'Hôpital, whose treatise
Analyse des infiniment petits, pour l'intelligences des lignes courbes (1696) was
a standard work on the differential calculus. L'Hôpital's first definition declares
"On appelle quantités *variables* celles qui augmentent ou diminuent continuellement;" and then continues: "La portion infiniment petite dont une quantité
variable augmente ou diminue continuellement, en est appellée la *Différence."*
His second postulate reads: "On demande qu'une ligne courbe puisse être con-
sidérée comme l'assemblage d'une infinité de lignes droites, chacune infiniment
petite: ou (ce qui est la même chose) comme un polygône d'un nombre infini
de côtes, chacun infiniment petit, lesquels déterminent par les angles qu'ils font
entr'eux la courbe de la ligne." (L'Hôpital 1696, 1-2)

never arrive at. And (which is most strange) although you should take a Million of Millions of these Infinitesimals, each whereof is supposed infinitely greater than some other real Magnitude, and add them to the least given Quantity, it shall be never the bigger. For this is one of the modest *postulata* of our modern Mathematicians, and is a Corner-stone or Ground-work of their Speculations.

§7. All these Points, I say, are supposed and believed by certain rigorous Exactors of Evidence in Religion, Men who pretend to believe no further than they can see. That Men, who have been conversant only about clear Points, should with difficulty admit obscure ones might not seem altogether unaccountable. But he who can digest a second or third Fluxion, a second or third Difference, need not, methinks, be squeamish about any Point in Divinity. There is a natural Presumption that Mens Faculties are made alike. It is on this Supposition that they attempt to argue and convince one another. What, therefore, shall appear evidently impossible and repugnant to one, may be presumed the same to another. But with what appearance of Reason shall any man presume to say, that Mysteries may not be Objects of Faith, at the same time that he himself admits such obscure Mysteries to be the Object of Science?

§8. It must indeed be acknowledged, the modern Mathematicians do not consider these Points as Mysteries, but as clearly conceived and mastered by their comprehensive Minds. They scruple not to say, that by help of these new Analytics they can penetrate into Infinity it self: That they can even extend their Views beyond Infinity: that their Art comprehends not only Infinite, but Infinite of Infinite (as they express it) or an Infinity of Infinities.[3] But, notwithstanding all these Assertions

---

[3] Here again, Berkeley is paraphrasing L'Hôpital. The Preface to *Analyse des infiniment petits* declares: "On peut même dire que cette Analyse s'étend au delà de l'infini: car elle ne se borne pas aux différences infiniment petites; mais elle découvre les rapports des différences de ces différences, ceux encore

and Pretensions, it may be justly questioned whether, as other Men in other Inquiries are often deceived by Words or Terms, so they likewise are not wonderfully deceived and deluded by their own peculiar Signs, Symbols, or Species.[4] Nothing is easier than to devise Expressions or Notations, for Fluxions and Infinitesimals of the first, second, third, fourth and subsequent Orders, proceeding in the same regular form without end or limit $\dot{x}$, $\ddot{x}$, $\dddot{x}$, $\ddddot{x}$, &c. or dx, ddx, $dddx$, $ddddx$, &c. These Expressions indeed are clear and distinct, and the Mind finds no difficulty in conceiving them to be continued beyond any assignable Bounds. But if we remove the Veil and look underneath, if laying aside the Expressions we set ourselves attentively to consider the things themselves, which are supposed to be expressed or marked thereby, we shall discover much Emptiness, Darkness, and Confusion; nay, if I mistake not, direct Impossibilities and Contradictions.[5] Whether this be the case or no, every thinking Reader is intreated to examine and judge for himself.

§9 Having considered the Object, I proceed to consider the Principles of this new Analysis by Momentums, Fluxions, or Infinitesimals; wherein if it shall appear that your capital Points, upon which the rest

---

de différences troisiémes, quatriémes, & ainsi de suite, sans trouver jamais de terme qui la puisse arrêter. De sorte qu'elle n'embrasse pas seulement l'infini; mais l'infini de l'infini, ou une infinité d'infinis." (L'Hôpital 1696, iii)

[4] The term "species" here is an antiquated term for what we today call variables. In Berkeley's day algebra was taken to be a kind of generalization of arithmetic in which letters stood for various kinds or "species" of magnitudes. Indeed, algebra itself was occasionally called "specious arithmetic," to indicate its relationship to ordinary arithmetic.

[5] Berkeley's insistence here upon "laying aside the Expressions" indicates an important difference between his account of algebra and his critique of the calculus. Berkeley explicitly endorses a strongly nominalistic reading of arithmetic and algebra in §§119-122 of the *Principles of Human Knowledge* and §§11-15 of Dialouge VII of the *Alciphron*, (*Works* 2, 95-97; *Works* 3, 303-309). Because he sees the calculus as a fundamentally geometric method, Berkeley rejects the possibility of justifying it on purely nominalistic grounds and demands that its key terms be interpreted in a sense consistent whith his reading of geometry as a science of extension. The distinction between algebra and the calculus is explored in Queries 27, 41, 45, and 46.

are supposed to depend, include Error and false Reasoning; it will then follow that you, who are at a loss to conduct your selves, cannot with any decency set up for guides to other Men. The main point in the Method of Fluxions is to obtain the Fluxion or Momentum of the Rectangle or Product of two indeterminate Quantities. Inasmuch as from thence are derived Rules for obtaining the Fluxions of all other Products and Powers; be the Coefficients or the Indexes what they will, integers or fractions, rational or surd.[6] Now this fundamental Point one would think should be very clearly made out, considering how much is built upon it, and that its Influence extends throughout the whole Analysis. But let the Reader judge. This is given for Demonstration.[*] Suppose the Product or Rectangle AB increased by continual Motion: and that the Momentaneous Increments of the Sides A and B are a and b. When the Sides A and B were deficient, or lesser by one half of their Moments, the Rectangle was $(A - \frac{1}{2}a) \times (B - \frac{1}{2}b)$ i.e. $AB - \frac{1}{2}aB - \frac{1}{2}bA + \frac{1}{4}ab$. And as soon as the Sides A and B are increased by the other two halves of their Moments, the Rectangle becomes $(A + \frac{1}{2}a) \times (B + \frac{1}{2}b)$ or $AB + \frac{1}{2}aB + \frac{1}{2}bA + \frac{1}{4}ab$. From the latter Rectangle subduct the former, and the remaining difference will be $aB + bA$. Therefore the Increment of the Rectangle generated by the intire Increments a and b is $aB + bA$. Q.E.D. But it is plain that the direct and true Method to obtain the Moment or Increment of the

---

[6] The term 'rectangle' in the previous sentence is an antiquated expression for 'product', reflecting the idea that (the area of) a rectangle is formed by the multiplication of its sides. Berkeley's point is that a method for determining the fluxion of a product can be extended to a method for finding the fluxion of a polynomial of any degree and with arbitrary coefficients. Once the fluxion of xy is found, powers such as x^2 are solved, and a general algorithm for finding the fluxion of a polynomial can be developed.

[*] Naturalis Philosophiae principia mathematica, l. 2 lem. 2. [In modern notation the following theorem is the "product rule" for differentiation. Given two functions $f(x)$ and $g(x)$ the derivative of the product $f(x)g(x)$ is $f'(x)g(x) + f(x)g'(x)$.]

Rectangle AB, is to take the Sides as increased by their whole Increments, and so multiply them together, $A + a$ by $B + b$, the Product
whereof $AB + aB + bA + ab$ is the augmented Rectangle; whence if we
subduct AB, the Remainder $aB+bA+ab$ will be the true Increment of the
Rectangle, exceeding that which was obtained by the former illegitimate
and indirect Method by the Quantity ab. And this holds universally be
the Quantities a and b what they will, big or little, Finite or Infinitesimal,
Increments, Moments, or Velocities. Nor will it avail to say that ab is a
Quantity exceeding small: Since we are told that *in rebus mathematicis
errores quàm minimi non sunt contemnendi**

§10. Such reasoning as this for Demonstration, nothing but the
obscurity of the Subject could have encouraged or induced the great
Author of the Fluxionary Method to put upon his Followers, and nothing
but an implicit deference to Authority could move them to admit. The
Case indeed is difficult. There can be nothing done till you have got rid
of the Quantity ab. In order to this the Notion of Fluxions is shifted:
It is placed in various Lights: Points which should be clear as first
Principles are puzzled; and Terms which should be steadily used are
ambiguous. But notwithstanding all this address and skill the point of
getting rid of ab cannot be obtained by legitimate reasoning. If a Man by
Methods, not geometrical or demonstrative, shall have satisfied himself
of the usefulness of certain Rules; which he afterwards shall propose to
his Disciples for undoubted Truths; which he undertakes to demonstrate
in a subtle manner, and by the help of nice and intricate Notions; it
is not hard to conceive that such his Disciples may, to save themselves
the trouble of thinking, be inclined to confound the usefulness of a Rule

---

* Intro. ad Quadraturam Curvaram. [Literally, "The most minute errors are
not in mathematical matters to be scorned." This slogan appears when Newton
insists that the ultimate ratios of vanishing quantities must be considered only
when the quantities have been diminished to nothing.]

with the certainty of a Truth, and accept the one for the other; especially if they are Men accustomed rather to compute than to think; earnest rather to go on fast and far, than solicitous to set out warily and see their way distinctly.

§11. The Points or mere Limits of nascent Lines are undoubtedly equal, as having no more Magnitude one than another, a Limit as such being no Quantity. If by a Momentum you mean more than the very initial Limit, it must be either a finite Quantity or an Infinitesimal. But all finite Quantities are expressly excluded from the Notion of a Momentum. Therefore the Momentum must be an Infinitesimal. And indeed, though much Artifice hath been employ'd to escape or avoid the admission of Quantities infinitely small, yet it seems ineffectual. For ought I see, you can admit no Quantity as a Medium between a finite Quantity and nothing, without admitting Infinitesimals. An Increment generated in a finite Particle of Time, is it self a finite Particle; and cannot therefore be a Momentum. You must therefore take an Infinitesimal Part of Time wherein to generate your Momentum. It is said, the Magnitude of Moments is not considered: And yet these same Moments are supposed to be divided into Parts. This is not easy to conceive, no more than it is why we should take Quantities less than A and B in order to obtain the Increment of AB, of which proceeding it must be owned the final Cause or Motive is very obvious; but it is not so obvious or easy to explain a just and legitimate Reason for it, or shew it to be Geometrical.

§12. From the foregoing Principle so demonstrated, the general Rule for finding the Fluxion of any Power of a flowing Quantity is derived.* But, as there seems to have been some inward Scruple or Consciousness of defect in the foregoing Demonstration, and as this finding the Fluxion of a given Power is a Point of primary Importance, it hath therefore been

---

* Philosophiae naturalis principia mathematica, lib. 2 lem. 2.

judged proper to demonstrate the same in a different manner independent of the foregoing Demonstration. But whether this other Method
be more legitimate and conclusive than the former, I proceed now to examine; and in order thereto shall premise the following Lemma. "If with
a View to demonstrate any Proposition, a certain Point is supposed, by
virtue of which certain other Points are attained; and such supposed
Point be it self afterwards destroyed or rejected by a contrary Supposition; in that case, all the other Points, attained thereby and consequent
thereupon, must also be destroyed and rejected, so as from thence forward to be no more supposed or applied in the Demonstration." This
is so plain as to need no Proof.

§13. Now the other Method of obtaining a Rule to find the Fluxion
of any Power is as follows.[7] Let the Quantity x flow uniformly, and be it
proposed to find the Fluxion of x^n. In the same time that x by flowing
becomes $x + o$, the Power x^n becomes $(x + o)^n$, i.e. by the Method of
infinite Series

$$x^n + nox^{n-1} + \frac{n^2 - n}{2}o^2x^{n-2} + \&c.,$$

and the Increments

$$o \quad \text{and} \quad nox^{n-1} + \frac{n^2 - n}{2}o^2x^{n-2} + \&c.,$$

are to one another as

$$1 \quad \text{to} \quad nx^{n-1} + \frac{n^2 - n}{2}ox^{n-2} + \&c.$$

Let now the Increments vanish, and their last Proportion will be 1 to
nx^{n-1}. But it should seem that this reasoning is not fair or conclusive.
For when it is said, let the Increments vanish, i.e. let the Increments be

---

[7] This demonstration is taken directly from the "Introduction" to Newton's
Quadrature of Curves. Although Berkeley does not acknowledge it, his presentation is a paraphrase of the Newtonian text in (*Papers*, 8: 127-9).

nothing, or let there be no Increments, the former Supposition that the Increments were something, or that there were Increments, is destroyed, and yet a Consequence of that Supposition, *i.e.* an Expression got by virtue thereof, is retained. Which, by the foregoing Lemma, is a false way of reasoning. Certainly when we suppose the Increments to vanish, we must suppose their Proportions, their Expressions, and every thing else derived from the Supposition of their Existence to vanish with them.

§14. To make this Point plainer, I shall unfold the reasoning, and propose it in a fuller light to your View. It amounts therefore to this, or may in other Words be thus expressed. I suppose that the Quantity x flows, and by flowing is increased, and its Increment I call o, so that by flowing it becomes $x+o$. And as x increaseth, it follows that every Power of x is likewise increased in a due Proportion. Therefore as x becomes $x+o$, x^n will become $(x+o)^n$: that is, according to the Method of infinite Series,

$$x^n + nox^{n-1} + \frac{n^2-n}{2}o^2x^{n-2} + \&c.$$

And if from the two augmented Quantities we subduct the Root and the Power respectively, we shall have remaining the two Increments, to wit,

$$o \quad \text{and} \quad nox^{n-1} + \frac{n^2-n}{2}o^2x^{n-2} + \&c.$$

which Increments, being both divided by the common Divisor o, yield the Quotients

$$1 \quad \text{and} \quad nx^{n-1} + \frac{n^2-n}{2}ox^{n-2} + \&c.$$

which are therefore Exponents of the Ratio of the Increments. Hitherto I have supposed that x flows, that x hath a real Increment, that o is something. And I have proceeded all along on that Supposition, without which I should not have been able to have made so much as one single

Step. From that Supposition it is that I get at the Increment of x^n, that I am able to compare it with the Increment of x, and that I find the Proportion between the two Increments. I now beg leave to make a new Supposition contrary to the first, *i.e.* I will suppose that there is no Increment of x, or that o is nothing; which second Supposition destroys my first, and is inconsistent with it, and therefore with every thing that supposeth it. I do nevertheless beg leave to retain nx^{n-1}, which is an Expression obtained in virtue of my first Supposition, which necessarily presupposeth such Supposition, and which could not be obtained without it: All which seems a most inconsistent way of arguing, and such as would not be allowed of in Divinity.

§15. Nothing is plainer than that no just Conclusion can be directly drawn from two inconsistent Suppositions. You may indeed suppose any thing possible: But afterwards you may not suppose any thing that destroys what you first supposed. Or if you do, you must begin *de novo*. If therefore you suppose that the Augments vanish, *i.e.* that there are no Augments, you are to begin again, and see what follows from such Supposition. But nothing will follow to your purpose. You cannot by that means ever arrive at your Conclusion, or succeed in, what is called by the celebrated Author, the Investigation of the first or last Proportions of nascent and evanescent Quantities, by Instituting the Analysis in finite ones.[8] I repeat it again: You are at liberty to

---

[8] The reference here is to Newton's remark at the end of his demonstration of the rule for determining the fluxion of any power of a flowing quantity. He writes: "By similar arguments there can by the method of first and last ratios be gathered the fluxions of lines, whether straight or curved, in any cases whatever, as also the fluxions of surface-areas, angles and other quantities. In finite quantities, however, to institute analysis in this way to investigate the first and last ratios of nascent or vanishing finites is in harmony with the geometry of the ancients, and I wanted to show that in the method of fluxions there should be no need to introduce infinitely small figures into geometry." (*Papers*, 8:129).

make any possible Supposition: And you may destroy one Supposition by another: But then you may not retain the Consequences, or any part of the Consequences of your first Supposition so destroyed. I admit that Signs may be made to denote either any thing or nothing: And consequently that in the original Notation $x + o$, o might have signified either an Increment or nothing. But then which of these soever you make it signify, you must argue consistently with such its Signification, and not proceed upon a double Meaning: Which to do were a manifest Sophism. Whether you argue in Symbols or in Words, the Rules of right Reason are still the same. Nor can it be supposed, you will plead a Privilege in Mathematics to be exempt from them.

§16. If you assume at first a Quantity increased by nothing, and in the Expression $x + o$, o stands for nothing, upon this Supposition as there is no Increment of the Root, so there will be no Increment of the Power; and consequently there will be none except the first, of all those Members of the Series constituting the Power of the Binomial; you will therefore never come at your Expression of a Fluxion legitimately by such Method. Hence you are driven into the fallacious way of proceeding to a certain Point on the Supposition of an Increment, and then at once shifting your Supposition to that of no Increment. There may seem great Skill in doing this at a certain Point or Period. Since if this second Supposition had been made before the common Division by o, all had vanished at once, and you must have got nothing by your Supposition. Whereas by this Artifice of first dividing, and then changing your Supposition, you retain 1 and nx^{n-1}. But, notwithstanding all this address to cover it, the fallacy is still the same. For whether it be done sooner or later, when once the second Supposition or Assumption is made, in the same instant the former Assumption and all that you got by it is destroyed, and goes out together. And this is universally true, be the Subject what it will, throughout all the Branches of humane Knowledge; in any other of which, I believe, Men would hardly admit such a reasoning as this, which in Mathematics is accepted for Demonstration.

§17. It may be not amiss to observe that the Method for finding the Fluxion of a Rectangle of two flowing Quantities, as it is set forth in the Treatise of Quadratures, differs from the abovementioned taken from the second Book of the Principles, and is in effect the same with that used in the *calculus differentialis*.* For the supposing a Quantity infinitely diminished, and therefore rejecting it, is in effect the rejecting an Infinitesimal; and indeed it requires a marvellous sharpness of Discernment, to be able to distinguish between evanescent Increments and infinitesimal Differences.[9] It may perhaps be said that the Quantity being infinitely diminished becomes nothing, and so nothing is rejected. But according to the received Principles it is evident, that no Geometrical Quantity, can by any division or subdivision whatsoever be exhausted, or reduced

---

* Analyse des infiniment petits, part 1. prop. 2. [L'Hôpital's text reads: "Proposition II. Problême: Prendre *la différence d'un produit fait de plusieurs quantités multipliés les unes par les autres*. 1°. La différence de *xy* est *ydx* + *xdy*. Car *y* devient *y* + *dy* lors que *x* devient *x* + *dx*; & partant *xy* devient alors *xy* + *ydx* + *xdy* + *dxdy*, qui est le produit de *x* + *dx* par *y* + *dy*, & sa différence sera *ydx* + *xdy* + *dxdy*; c'est à dire *ydx* + *xdy*: puisque *dxdy* est une quantité infiniment petite par rapport aux autres termes *ydx*, & *xdy*" This first case is then extended to successive multiplications, always rejecting higher-order infinitesimals to obtain the desired result.]

[9] Berkeley is here denying a common claim advanced by proponents of Newton in his priority dispute with Leibniz over the invention of the calculus. Partisans of Newton argued that, not only had Leibniz plagiarized the calculus, but he had also introduced changes which made the method unrigorous. In particular, the Newtonian method of prime and ultimate ratios was frequently touted as a rigorous alternative to the *calculus differentialis*. In "An Account of a Book Entitled *Commercium Epistolicum*. . ." in the *Philosophical Transactions*, Newton (through his spokesman Keill) declares that "[The Method of Fluxions] is more Natural and Geometrical, because founded on the *primæ quantitatum nascentium rationes*, which have a Being in Geometry, while *Indivisibles*, upon which the Differential Method is founded, have no Being in Geometry or in Nature,"(Hall 1980, 295). See Hall (1980) for a study of this controversy and a reprint of Newton's "Account."

to nothing.[10]   Considering the various Arts and Devices used by the
great Author of the Fluxionary Method: in how many Lights he placeth
his Fluxions: and in what different ways he attempts to demonstrate the
same Point: one would be inclined to think, he was himself suspicious
of the justness of his own demonstrations; and that he was not enough
pleased with any one notion steadily to adhere to it. This much at
least is plain, that he owned himself satisfied concerning certain Points,
which nevertheless he could not undertake to demonstrate to others.*
Whether this Satisfaction arose from tentative Methods or Inductions;
which have often been admitted by Mathematicians (for instance by Dr.
Wallis in his Arithmetic of Infinities)[11] is what I shall not pretend to
determine. But, whatever the Case might have been with respect to the
Author, it appears that his Followers have shewn themselves more eager
in applying his Method, than accurate in examining his Principles.

---

[10] Here, Berkeley comes close to endorsing the thesis of infinite divisibility,
which he elsewhere denies (*Cf.* Queries 5, 6, 18 and 52).

* *See Letter to Collins,* Nov. 8, 1676. [Berkeley's reference is to an extract of
a letter from Newton to John Collins, published as part of the Royal Society's
Commercium Epistolicum D. Johannis Collins, et Aliorum de Analysi Promota
(London: 1712/3) as part of Newton's claim against Leibniz in the priority
dispute over the discovery of the calculus. The key passage which interests
Berkeley reads "I say there is no such curve line, but I can, in less than half a
quarter of an hour, tell whether it may be squared, or what are the simplest
figures it may be compared with, be those figures conic sections or others....
This may seem a bold assertion, because it is hard to say a figure may or may
not be squared or compared with another, but it is plain to me by the fountain I
draw it from, though I will not undertake to prove it to others." (Rigaud [1841]
1965, 2: 404)]

[11] The reference here is to Wallis's casual way of "proving" arithmetical results in the summation of infinite series by investigating a few initial cases
and concluding "by induction" that the result holds in the infinite case. A
specimen of Wallis's procedure is reproduced in Section 1.2 of the "Editor's
Introduction." His attitude is expressed in Proposition I of his *Arithmetica
Infinitorum*: "Simplicissimus investigandi modus, in hoc & sequentibus aliquot
Problematis, est, rem ipsam aliquousque præstare, & rationes producentes ob-
servare atque invicem comparare; ut inductione tandem universalis propositio
innotescat." (Wallis 1693-99, 1: 365.)

§18. It is curious to observe, what subtilty and skill this great Genius employs to struggle with an insuperable Difficulty; and through what Labyrinths he endeavours to escape the doctrine of Infinitesimals; which as it intrudes upon him whether he will or no, so it is admitted and embraced by others without the least repugnance. *Leibnitz* and his Followers in their *calculus differentialis* making no manner of Scruple, first to suppose, and secondly to reject Quantities infinitely small: with what clearness in the Apprehension and justness in the Reasoning, any thinking Man, who is not prejudiced in favour of those things, may easily discern. The Notion or Idea of an infinitesimal Quantity, as it is an Object simply apprehended by the Mind, hath been already considered.* I shall now only observe as to the method of getting rid of such Quantities, that it is done without the least Ceremony. As in Fluxions the Point of first importance, and which paves the way to the rest, is to find the Fluxion of a Product of two indeterminate Quantities, so in the *calculus differentialis* (which Method is supposed to have been borrowed from the former with some small Alterations)[12] the main Point is to obtain the difference of such Product. Now the Rule for this is got by rejecting the Product or Rectangle of the Differences. And in general it is supposed, that no Quantity is bigger or lesser for the Addition or Subduction of its Infinitesimal: and that consequently no error can arise from such rejection of Infinitesimals.

§19. And yet it should seem that, whatever errors are admitted in the Premises, proportional errors ought to be apprehended in the Conclusion, be they finite or infinitesimal: and that therefore the ἀκρίβεια of

---

* *Sect. 5 and 6.*

[12] Another reference to the famous priority dispute between Newton and Leibniz.

Geometry requires nothing should be neglected or rejected.[13] In answer
to this you will perhaps say, that the Conclusions are accurately true,
and that therefore the Principles and Methods from whence they are
derived must be so too. But this inverted way of demonstrating your
Principles by your Conclusions, as it would be peculiar to you Gentlemen, so it is contrary to the Rules of Logic. The truth of the Conclusion
will not prove either the Form or the Matter of a Syllogism to be true:
inasmuch as the Illation might have been wrong or the Premises false,
and the Conclusion nevertheless true, though not in virtue of such Illation or of such Premises.[14] I say that in every other Science Men prove
their Conclusions by their Principles, and not their Principles by the
Conclusions. But if in yours you should allow your selves this unnatural
way of proceeding, the Consequence would be that you must take up
with Induction, and bid adieu to Demonstration. And if you submit to
this, your Authority will no longer lead the way in Points of Reason and
Science.

§20. I have no Controversy about your Conclusions, but only about
your Logic and Method. How you demonstrate? What Objects you
are conversant with, and whether you conceive them clearly? What
Principles you proceed upon; how sound they may be; and how you apply
them? It must be remembered that I am not concerned about the truth

---

[13] The term '$\dot{\alpha}\kappa\rho\dot{\iota}\beta\epsilon\iota\alpha$' is the Greek for exactness or precision. Berkeley
uses it in entry 313 of his *Philosophical Commentaries*, writing "What shall
I say? dare I pronounce the admir'd $\dot{\alpha}\kappa\rho\dot{\iota}\beta\epsilon\iota\alpha$ Mathematica, that Darling of
the Age a trifle?" (*Works*, 1: 39). More interestingly, Brook Taylor uses it
extoll the virtues of the Newtonian calculus of fluxions over infinitesimal approaches: "Cavallerius & Recentiores contemplarunt partes istas ut in infinitum
diminutas. Sed hi omnes, contemplando geneses quantitatum per additiones
partium, non satis consuluerunt severæ isti $\dot{\alpha}\kappa\rho\dot{\iota}\beta\epsilon\iota\alpha$ Geometrarum."(Taylor
1715, Preface)

[14] The term 'illation' here is an antiquated term for inference. Berkeley's
point is that an argument with a true conclusion can still have either an invalid
formal structure or a false premise, so the truth of the conclusion does not
guarantee the validity of the form or the truth of the premises.

of your Theorems, but only about the way of coming at them; whether
it be legitimate or illegitimate, clear or obscure, scientific or tentative.
To prevent all possibility of your mistaking me, I beg leave to repeat and
insist, that I consider the Geometrical Analyst only as a Logician, *i.e.*
so far forth as he reasons and argues; and his Mathematical Conclusions,
not in themselves, but in their Premises; not as true or false, useful or
insignificant, but as derived from such Principles, and by such Inferences.
And forasmuch as it may perhaps seem an unaccountable Paradox, that
Mathematicians should deduce true Propositions from false Principles,
be right in the Conclusion and yet err in the Premises; I shall endeavour
particularly to explain why this may come to pass, and shew how Error
may bring forth Truth, though it cannot bring forth Science.

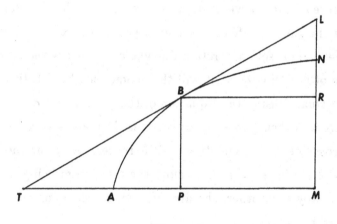

Figure 1

§21. In order therefore to clear up this Point, we will suppose for
instance that a Tangent is to be drawn to a Parabola, and examine the
progress of this Affair, as it is performed by infinitesimal Differences. Let
AB be a Curve, the Abscisse $AP = x$, the ordinate $PB = y$, the Difference

of the Abscisse $PM = dx$, the Difference of the ordinate $RN = dy$ [Figure 1]. Now by supposing the Curve to be a Polygon, and consequently BN, the Increment or Difference of the Curve, to be a straight Line coincident with the Tangent, and the differential Triangle BRN to be similar to the triangle TPB the Subtangent PT is found a fourth Proportional to $RN : RB : PB$: that is to $dy : dx : y$. Hence the Subtangent will be $\frac{y\,dx}{dy}$. But herein there is an error arising from the forementioned false supposition, whence the value PT comes out greater than the Truth: for in reality it is not the Triangle RNB but RLB, which is similar to PBT, and therefore (instead of RN) RL should have been the first term of the Proportion, *i.e.* $RN + NL$, *i.e.* $dy + z$: whence the true expression for the Subtangent should have been $\frac{y\,dx}{dy+z}$. There was therefore an error of defect in making dy the divisor: which error was equal to z, *i.e.* NL the Line comprehended between the curve and the Tangent. Now by the nature of the Curve $y^2 = px$, supposing p to be the Parameter, whence by the rule of Differences $2y\,dy = p\,dx$ and $dy = \frac{p\,dx}{2y}$. But if you multiply $y+dy$ by it self, and retain the whole Product without rejecting the Square of the Difference, it will then come out, by substituting the augmented Quantities in the Equation of the Curve, that $dy = \frac{p\,dx}{2y} - \frac{dy^2}{2y}$ truly. There was therefore an error of excess in making $dy = \frac{p\,dx}{2y}$, which followed from the erroneous Rule of Differences. And the measure of this second error is $\frac{dy^2}{2y} = z$. Therefore the two errors being equal and contrary destroy each other; the first error of defect being corrected by a second error of excess.[15]

---

[15] Berkeley's strategy in this argument for compensating errors can be unfolded as follows: he first finds an expression for the subtangent PT in terms of y, dy, and dx; by construction, $RN : RB :: PB : PT$, so $PT = \frac{y\,dx}{dy}$. Then he seeks a convenient expression for dy which can later be used to eliminate dx and dy. This he gets by taking the derivative of the equation $y^2 = px$, which is $2y\,dy = p\,dx$, so that $dy = \frac{p\,dx}{2y}$. Substituting this value for dy in the equation for PT, we get $PT = \frac{2y^2}{p}$, and the problem is solved. To show the

§22. If you had committed only one error, you would not have come at a true Solution of the Problem. But by virtue of a twofold mistake you arrive, though not at Science, yet at Truth. For Science it cannot be called, when you proceed blindfold, and arrive at the Truth not knowing how or by what means. To demonstrate that z is equal to $\frac{dy^2}{2y}$, let BR or dx be m, and RN or dy be n. By the thirty third Proposition of the first Book of the Conics of *Apollonius*,[16] and from similar Triangles, as $2x$ to y so is m to $n + z = \frac{my}{2x}$. Likewise from the Nature of the Parabola $y^2 + 2yn + n^2 = xp + mp$, and $2yn + n^2 = mp$: wherefore $\frac{2yn+n^2}{p} = m$: and because $y^2 = px$, $\frac{y^2}{p}$ will be equal to x. Therefore substituting these values instead of m and x we shall have

$$n + z = \frac{my}{2x} = \frac{2y^2 np + yn^2 p}{2y^2 p} \quad :$$

i.e.

$$n + z = \frac{2yn + n^2}{2y} \quad :$$

which being reduced gives

$$z = \frac{n^2}{2y} = \frac{dy^2}{2y} \quad Q.E.D.$$

---

compensating errors, he then substitutes the neglected quantities z and dy^2 into the reasoning. The first of these was neglected when the curve was treated as a polygon, the second when a higher-order differential was discarded from the expression for the derivative. Now the task is to show that the errors are equal and opposite, which he undertakes in the next section by arguing that $z = \frac{dy^2}{2y}$, on the basis of results from the theory of conic sections.

[16] The theorem invoked here is the "tangent-axis theorem" which states, in essence, that the subtangent to the parabola is bisected at the vertex. The Apollonian statement of the theorem reads; "If in a parabola some point is taken, and from it an ordinate is dropped to the diameter, and, to the straight line cut off by it on the diameter from the vertex, a straight line in the same straight line from its extremity is made equal, then the straight line joined from the point thus resulting to the point taken will touch the section," (Apollonius of Perga [1939] 1952, 640).

§23. Now I observe in the first place, that the Conclusion comes out right, not because the rejected Square of *dy* was infinitely small; but because this error was compensated by another contrary and equal error. I observe in the second place, that whatever is rejected, be it ever so small, if it be real and consequently makes a real error in the Premises, it will produce a proportional real error in the Conclusion. Your Theorems therefore cannot be accurately true, nor your Problems accurately solved, in virtue of Premises, which themselves are not accurate, it being a rule in Logic that *Conclusio sequitur partem debiliorem*.[17] Therefore I observe in the third place, that when the Conclusion is evident and the Premises obscure, or the Conclusion accurate and the Premises inaccurate, we may safely pronounce that such Conclusion is neither evident nor accurate, in virtue of those obscure inaccurate Premises or Principles; but in virtue of some other Principles which perhaps the Demonstrator himself never knew or thought of. I observe in the last place, that in case the Differences are supposed finite Quantities ever so great, the Conclusion will nevertheless come out the same: inasmuch as the rejected Quantities are legitimately thrown out, not for their smallness, but for another reason, to wit, because of contrary errors, which destroying each other do upon the whole cause that nothing is really, though something is apparently thrown out. And this Reason holds equally, with respect to Quantities finite as well as infinitesimal, great as well as small, a Foot or a Yard long as well as the minutest Increment.[18]

---

[17] Literally, "the conclusion follows the weaker part." The idea is that an inference is no more reliable than its most questionable premise.

[18] The significance of this argument for the compensation of errors thesis is perhaps less than Berkeley imagined, but he has succeded in showing that the methods of the calculus can be avoided in some cases and replaced by classical results from the theory of conic sections. Grattan-Guinness (1969, 223) has argued that Berkeley's argument "may be generalised to any function susceptible of expansion by a Taylor series," but Berkeley has no argument to show compensating errors in every application of the calculus. Breidert (1989, 103)

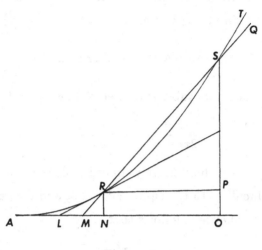

Figure 2

§24. For the fuller illustration of this Point, I shall consider it in another light, and proceeding in finite Quantities to the Conclusion, I shall only then make use of one Infinitesimal. Suppose the straight Line MQ cuts the Curve AT in the Points R and S [Figure 2]. Suppose LR a Tangent at the Point R, AN the Abscisse, NR and OS Ordinates. Let AN be produced to O, and RP be drawn parallel to NO. Suppose $AN = x$, $NR = y$, $NO = v$, $PS = z$, the subsecant $MN = s$.[19] Let th⌐ Equation $y = x^2$ express the nature of the Curve: and supposing y and x increased by their finite Increments we get

$$y + z = x^2 + 2xv + v^2 :$$

observes "Berkeley geht z.B. nicht auf die Frage ein. . . , ob sich das unterstellte Verfahren der Fehlerkompensation, das er am Beispiel der Parabel vorrechnete, auch allgemein in jedem Falle durchführen lasse." The idea of basing the calculus on compensating errors did not end with Berkeley; the most complete attempt at such a program was developed by Lazare Carnot in his *Réflexions sur la métaphysique du calcul infinitésimal*. See Grabiner (1981) for more on this topic.
[19] Reading 's' for Berkeley's 'S' to retain notational consistency.

whence the former Equation being subducted there remains $z = 2xv + v^2$. And by reason of similar Triangles

$$PS : PR :: NR : NM, \quad i.e. \quad z : v :: y : s = \frac{vy}{z},$$

wherein if for y and z we substitute their values, we get

$$\frac{vx^2}{2xv + v^2} = s = \frac{x^2}{2x + v}.$$

And supposing NO to be infinitely diminished, the subsecant NM will in that case coincide with the subtangent NL, and v as an Infinitesimal may be rejected, whence it follows that

$$s = NL = \frac{x^2}{2x} = \frac{x}{2} \quad ;$$

which is the true value of the Subtangent. And since this was obtained by one only error, *i.e.* by once rejecting one only Infinitesimal, it should seem, contrary to what hath been said, that an infinitesimal Quantity or Difference may be neglected or thrown away, and the Conclusion nevertheless be accurately true, although there was no double mistake or rectifying of one error by another, as in the first Case. But if this Point be thoroughly considered, we shall find there is even here a double mistake, and that one compensates or rectifies the other. For in the first place, it was supposed, that when NO is infinitely diminished or becomes an Infinitesimal then the Subsecant NM becomes equal to the Subtangent NL. But this is a plain mistake, for it is evident, that as a Secant cannot become a Tangent, so a Subsecant cannot be a Subtangent. Be the Difference ever so small, yet still there is a Difference. And if NO be infinitely small, there will even then be an infinitely small Difference between NM and NL. Therefore NM or S was too little for your supposition, (when you supposed it equal to NL) and this error was compensated by a second error in throwing out v, which last error made s bigger than its true value, and in lieu thereof gave the value of the Subtangent. This is the true State of the Case, however it may be disguised. And to this in reality it amounts, and is at bottom the same thing, if we should pretend to find the Subtangent by having first found, from the Equation of the Curve and similar Triangles, a general Expression for all Subsecants, and then reducing the Subtangent under this general Rule, by considering it as the Subsecant when v vanishes or becomes nothing.[20]

§25. Upon the whole I observe, *First,* that v can never be nothing so long as there is a secant. *Secondly,* that the same Line cannot be both tangent and secant. *Thirdly,* that when v or NO^* vanisheth, PS and SR do also vanish, and with them the proportionality of the similar Triangles. Consequently the whole Expression, which was obtained by means thereof and grounded thereupon, vanisheth when v vanisheth. *Fourthly,* that the Method for finding Secants or the Expression of Secants, be it ever so general, cannot in common sense extend any further than to all Secants whatsoever: and, as it necessarily supposeth similar Triangles, it cannot be supposed to take place where there are not similar Triangles. *Fifthly,* that the Subsecant will always be less than the Subtangent, and can never coincide with it; which Coincidence to suppose would be absurd; for it would be supposing the same Line at the same

---

[20] The reasoning in this argument for compensating errors is both less complicated and less convincing than in the previous case. The basic idea is that the first "error" of neglecting v is compensated when we assume that NM coincides with NL; Berkeley asserts that this is impossible, because a secant must cut the curve in two points and the tangent merely touch the curve at one point. On this basis he argues that there must therefore be some difference between the secant and tangent. He then assumes without argument that this difference $NL - NM$ is equal to the neglected v or NO. But the assumption is groundless, and this particular argument for compensating errors falls short of proof.

* *See the foregoing figure.*

time to cut and not to cut another given Line, which is a manifest Contradiction, such as subverts the Hypothesis and gives a Demonstration of its Falshood. *Sixthly*, if this be not admitted, I demand a Reason why any other apagogical Demonstration,[21] or Demonstration *ad absurdum* should be admitted in Geometry rather than this: Or that some real Difference be assigned between this and others as such. *Seventhly*, I observe that it is sophistical to suppose NO or RP, PS, and SR to be finite real Lines in order to form the Triangle, RPS, in order to obtain Proportions by similar Triangles; and afterwards to suppose there are no such Lines, nor consequently similar Triangles, and nevertheless to retain the Consequence of the first Supposition, after such Supposition hath been destroyed by a contrary one. *Eighthly*, That although, in the present case, by inconsistent Suppositions Truth may be obtained, yet that such Truth is not demonstrated: That such Method is not conformable to the Rules of Logic and right Reason: That, however useful it may be, it must be considered only as a Presumption, as a Knack, an Art or rather an Artifice, but not a scientific Demonstration.

§26. The Doctrine premised may be farther illustrated by the following simple and easy Case, wherein I shall proceed by evanescent Increments. Suppose $AB = x$, $BC = y$, $BD = o$, and that x^2 is equal to the Area ABC: it is proposed to find the ordinate y or BC [Figure 3]. When x by flowing becomes $x + o$, then x^2 becomes $x^2 + 2xo + o^2$: And the Area ABC becomes ADH, and the Increment of x^2 will be equal to

---

[21] The term 'apagogical' derives from the Greek ἀπαγωγή, meaning reduction. Berkeley's point is that he takes the foregoing demonstration to have been reduced to the absurdity of claiming that a line both cuts and does not cut the circle. His claim is not terribly compelling, however, because the continued rotation of a secant about one of the points in which it cuts a curve will produce a tangent.

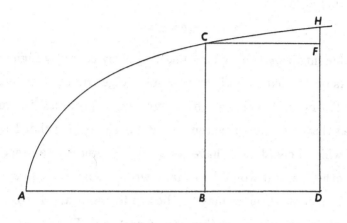

Figure 3

$BDHC$ the Increment of the Area, *i.e.* to $BCFD + CFH$. And if we suppose the curvilinear space CFH to be qo^2, then

$$2xo + o^2 = yo + qo^2$$

which divided by o gives $2x + o = y + qo$. And, supposing o to vanish, $2x = y$, in which Case ACH will be a straight Line, and the Areas ABC, CFH, Triangles. Now with regard to this Reasoning, it hath been already remarked,* that it is not legitimate or logical to suppose o to vanish, *i.e.* to be nothing, *i.e.* that there is no Increment, unless we reject at the same time with the Increment it self every Consequence of such Increment, *i.e.* whatsoever could not be obtained but by supposing such Increment. It must nevertheless be acknowledged that the Problem is rightly solved, and the Conclusion true, to which we are led by this Method. It will therefore be asked, how comes it to pass that the throwing out o is attended with no Error in the Conclusion? I answer,

---

* *Sect.* 12 *and* 13 supra.

the true reason hereof is plainly this: Because q being Unite, qo is equal
to o: and therefore

$$2x + o - qo = y = 2x,$$

the equal Quantities qo and o being destroyed by contrary Signs.[22]

§27. As on the one hand it were absurd to get rid of o by saying, let
me contradict my self: Let me subvert my own Hypothesis: Let me take
for granted that there is no Increment, at the same time that I retain a
Quantity, which I could never have got at but by assuming an Increment:
So on the other hand it would be equally wrong to imagine, that in a geometrical Demonstration we may be allowed to admit any Error, though
ever so small, or that it is possible, in the nature of Things, an accurate
Conclusion should be derived from inaccurate Principles. Therefore o
cannot be thrown out as an Infinitesimal, or upon the Principle that
Infinitesimals may be safely neglected. But only because it is destroyed
by an equal Quantity with a negative Sign, whence $o - qo$ is equal to
nothing. And as it is illegitimate to reduce an Equation, by subducting
from one Side a Quantity when it is not to be destroyed, or when an
equal Quantity is not subducted from the other Side of the Equation: So
it must be allowed a very logical and just Method of arguing, to conclude
that if from Equals either nothing or equal Quantities are subducted,
they shall still remain equal. And this is the true Reason why no Error
is at last produced by the rejecting of o. Which therefore must not be
ascribed to the Doctrine of Differences, or Infinitesimals, or evanescent
Quantities, or Momentums, or Fluxions.

---

[22] The two key assumptions here are that $CFH = qo^2$ and $q = 1$. Although it
is certainly possible in any given case to assign a number q such that $CFH = qo^2$, q will not generally be equal to 1, and the argument for compensating
errors breaks down in this case. See Wisdom (1941, 59-61) and Breidert (1989,
104-5) for other analyses of this argument and its weaknesses.

§28. Suppose the Case to be general, and that x^n is equal to the Area ABC, whence by the Method of Fluxions the Ordinate is found nx^{n-1} which we admit for true, and shall inquire how it is arrived at. Now if we are content to come at the Conclusion in a summary way, by supposing that the Ratio of the Fluxions of x and x^n are found* to be 1 and nx^{n-1}, and that the Ordinate of the Area is considered as its Fluxion; we shall not so clearly see our way, or perceive how the truth comes out, that Method as we have shewed before being obscure and illogical. But if we fairly delineate the Area and its Increment, and divide the latter into two Parts $BCFD$ and CFH[†], and proceed regularly by Equations between the algebraical and geometrical Quantities, the reason of the thing will plainly appear. For as x^n is equal to the Area ABC, so is the Increment of x^n equal to the Increment of the Area, i.e. to $BDHC$; that is, to say,

$$no x^{n-1} + \frac{n^2 - n}{2} o^2 x^{n-2} + \&c. = BDFC + CFH.$$

And only the first Member on each Side of the Equation being retained, $no x^{n-1} = BDFC$: And dividing both Sides by o or BD, we shall get $nx^{n-1} = BC$. Admitting, therefore, that the curvilinear Space CFH is equal to the rejectaneous Quantity

$$\frac{n^2 - n}{2} o^2 x^{n-2} + \&c.$$

and that when this is rejected on one Side, that is rejected on the other, the Reasoning becomes just and the Conclusion true.[23] And it is all one

---

* *Sect. 13*

† *See the figure in Sect. 26.*

[23] Again, Berkeley's case for compensating errors falls victim to an unwarranted assumption. It is certainly true that the two sums are equal, but it by no means follows from this that the *first term* in each of the two sums must be equal. In short, Berkeley has illegitimately moved from a given equation of the form $\alpha + \beta = \gamma + \delta$ to the conclusion that $\alpha = \gamma$ and $\beta = \delta$.

whatever Magnitude you allow to BD, whether that of an infinitesimal Difference or a finite Increment ever so great. It is therefore plain, that the supposing the rejectaneous algebraical Quantity to be an infinitely small or evanescent Quantity, and therefore to be neglected, must have produced an Error, had it not been for the curvilinear Spaces being equal thereto, and at the same time subducted from the other Part or Side of the Equation agreeably to the Axiom, *If from Equals you subduct Equals, the Remainders will be equal.* For those Quantities which by the Analysts are said to be neglected, or made to vanish, are in reality subducted. If therefore the Conclusion be true, it is absolutely necessary that the finite space CFH be equal to the Remainder of the Increment expressed by

$$\frac{n^2 - n}{2} o^2 x^{n-2} + \&c.$$

equal I say to the finite Remainder of a finite Increment.[24]

§29. Therefore, be the Power what you please, there will arise on one Side an algebraical Expression, on the other a geometrical Quantity, each of which naturally divides it self into three Members: The algebraical or fluxionary Expression, into one which includes neither the Expression of the Increment of the Absciss nor of any Power thereof, another which includes the Expression of the Increment itself, and a third including the Expression of the Powers of the Increment. The geometrical Quantity also or whole increased Area consists of three Parts or Members, the first of which is the given Area, the second a Rectangle under the Ordinate and the Increment of the Absciss, and the third a curvilinear Space. And, comparing the homologous or correspondent Members on

---

[24] Here, Berkeley falls into exactly the kind of illegitimate argument he had earlier denounced in §19. He defends the questionable assumption that had led to his conclusion by citing the truth of the conclusion as evidence for the truth of the assumption.

both Sides, we find that as the first Member of the Expression is the Expression of the given Area, so the second Member of the Expression will express the Rectangle or second Member of the geometrical Quantity; and the third, containing the Powers of the Increment, will express the curvilinear Space, or third Member of the geometrical Quantity. This hint may, perhaps, be further extended and applied to good purpose, by those who have leisure and curiosity for such Matters. The use I make of it is to shew, that the Analysis cannot obtain in Augments or Differences, but it must also obtain in finite Quantities, be they ever so great, as was before observed.[25]

§30. It seems therefore upon the whole that we may safely pronounce, the Conclusion cannot be right, if in order thereto any Quantity be made to vanish, or be neglected, except that either one Error is redressed by another; or that secondly, on the same Side of an Equation equal Quantities are destroyed by contrary Signs, so that the Quantity we mean to reject is first annihilated; or lastly, that from the opposite Sides equal Quantities are subducted. And therefore to get rid of Quantities by the received Principles of Fluxions or of Differences is neither good Geometry nor good Logic. When the Augments vanish, the Velocities also vanish. The Velocities or Fluxions are said to be *primò* or *ultimò*, as the Augments nascent and evanescent. Take therefore the *Ratio* of the evanescent Quantities, it is the same with that of the Fluxions. It will therefore answer all Intents as well. Why then are Fluxions introduced? Is it not to shun or rather to palliate the Use of Quantities infinitely small? But we have no Notion whereby to conceive and measure various Degrees of Velocity, beside Space and Time, or when the Times are given, beside Space alone. We have even no Notion of

---

[25] Berkeley refers back to the two previous sections in Query 37. He there suggests that the compensation of errors thesis could be exploited to develop an alternative to the calculus within the confines of classical geometry.

Velocity prescinded from Time and Space. When therefore a Point is supposed to move in given Times, we have no Notion of greater or lesser Velocities or of Proportions between Velocities, but only of longer or shorter Lines, and of Proportions between such Lines generated in equal Parts of Time.

§31. A Point may be the limit of a Line: A Line may be the limit of a Surface: A Moment may terminate Time.[26] But how can we conceive a Velocity by the help of such Limits? It necessarily implies both Time and Space, and cannot be conceived without them. And if the Velocities of nascent and evanescent Quantities, *i.e.* abstracted from Time and Space, may not be comprehended, how can we comprehend and demonstrate their Proportions? Or consider their *rationes primæ* and *ultimæ*? For to consider the Proportion or *Ratio* of Things implies that such Things have Magnitude: That such their Magnitudes may be measured, and their Relations to each other known. But, as there is no measure of Velocity except Time and Space, the proportion of Velocities being only compounded of the direct Proportion of the Spaces, and the reciprocal Proportion of the Times; doth it not follow that to talk of investigating, obtaining, and considering the Proportions of Velocities, exclusively of Time and Space, is to talk unintelligibly?[27]

§32. But you will say that, in the use and application of Flux-ions, Men do not overstrain their Faculties to a precise Conception of

---

[26] Berkeley is here using the term 'limit' as equivalent to 'extremity' or 'boundary'. His point of reference is definitions 3 and 6 in Book I of Euclid's *Elements*: "3. The extremities of a line are points. . . . 6. The extremeties of a surface are lines," (Euclid [1925] 1956, 1: 153). He is not using the term 'limit' in the modern sense.

[27] The critique of the calculus in this and the foregoing section recalls Berkeley's famous attacks on abstract ideas in the "Introduction" to his *Principles of Human Knowledge* and other writings. He is here directing his attention to a characterization of fluxions which would define them as instantaneous velocities, abstracted from any consideration of time and space. Similar points are raised in §37, §44, and Queries 29-30.

the abovementioned Velocities, Increments, Infinitesimals, or any other
such like Ideas of a Nature so nice, subtile, and evanescent. And therefore you will perhaps maintain, that Problems may be solved without
those inconceivable Suppositions: and that, consequently, the Doctrine
of Fluxions, as to the practical Part, stands clear of all such Difficulties.
I answer, that if in the use or application of this Method, those difficult
and obscure Points are not attended to, they are nevertheless supposed.
They are the Foundation on which the Moderns build, the Principles on
which they proceed, in solving Problems and discovering Theorems. It
is with the Method of Fluxions as with all other Methods, which pre-
suppose their respective Principles and are grounded thereon. Although
the Rules may be practised by Men who neither attend to, nor perhaps
know the Principles. In like manner, therefore, as a Sailor may prac-
tically apply certain Rules derived from Astronomy and Geometry, the
Principles whereof he doth not understand: And as any ordinary Man
may solve divers numerical Questions, by the vulgar Rules and Operations of Arithmetic, which he performs and applies without knowing
the Reasons of them: Even so it cannot be denied that you may apply the Rules of the fluxionary Method: You may compare and reduce
particular Cases to general Forms: You may operate and compute and
solve Problems thereby, not only without an actual Attention to, or an
actual Knowledge of, the Grounds of that Method, and the Principles
whereon it depends, and whence it is deduced, but even without having
ever considered or comprehended them.

§33. But then it must be remembered, that in such Case although
you may pass for an Artist, Computist, or Analyst, yet you may not be
justly esteemed a Man of Science and Demonstration.[28] Nor should any

---

[28] Berkeley's insistence here upon the necessity of comprehending the fundamental principles of the calculus undercuts interpretations which portray him as
an instrumentalist in the philosophy of mathematics. Baum (1972) has argued

Man, in virtue of being conversant in such obscure Analytics, imagine
his rational Faculties to be more improved than those of other Men,
which have been exercised in a different manner, and on different Subjects; much less erect himself into a Judge and an Oracle, concerning
Matters that have no sort of connexion with, or dependence on those
Species, Symbols or Signs, in the Management whereof he is so conversant and expert. As you, who are a skillful Computist or Analyst, may
not therefore be deemed skillful in Anatomy: or *vice versa*, as a Man
who can dissect with Art, may, nevertheless, be ignorant of your Art of
computing: Even so you may both, notwithstanding your peculiar Skill
in your respective Arts, be alike unqualified to decide upon Logic, or
Metaphysics, or Ethics, or Religion. And this would be true, even admitting that you understood your own Principles and could demonstrate
them.

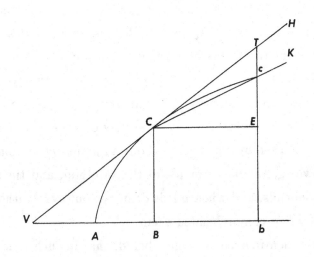

Figure 4

for an instrumentalistic and formalistic interpretation of Berkeley's philosophy
of mathematics, but has overlooked *The Analyst*.

§34. If it is said, that Fluxions may be expounded or expressed
by finite Lines proportional to them: Which finite Lines, as they may
be distinctly conceived and known and reasoned upon, so they may be
substituted for the Fluxions, and their mutual Relations or Proportions
be considered as the Proportions of Fluxions: By which means the Doctrine becomes clear and useful.[29] I answer that if, in order to arrive at
these finite Lines proportional to the Fluxions, there be certain Steps
made use of which are obscure and inconceivable, be those finite Lines
themselves ever so clearly conceived, it must nevertheless be acknowledged, that your proceeding is not clear nor your method scientific. For
instance, it is supposed [in Figure 4] that AB being the Absciss, BC the
Ordinate, and VCH a Tangent of the Curve AC, Bb or CE the Increment of the Absciss, Ec the Increment of the Ordinate, which produced
meets VH in the Point T, and Cc the Increment of the Curve. The right
Line Cc being produced to K, there are formed three small Triangles,
the Rectilinear CEc, the Mixtilinear CEc, and the Rectilinear Triangle
CET. It is evident these three Triangles are different from each other,
the Rectilinear CEc being less than the Mixtilinear CEc, whose Sides
are the three Increments abovementioned, and this still less than the
Triangle CET. It is supposed that the Ordinate bc moves into the place
BC, so that the Point c is coincident with the Point C; and the right line
CK, and consequently the Curve Cc, is coincident with the Tangent CH.
In which case the mixtilinear evanescent Triangle CEc will, in its last
form, be similar to the Triangle CET: And its evanescent Sides CE, Ec,
and Cc will be proportional to CE, ET, and CT the Sides of the Triangle

---

[29] The reference here is to Newton's *Quadrature of Curves*, where Newton
proposes to replace proportions of fluxions by proportions of the sides of a
triangle which stand to one another in the same ratio as the nascent or evanescent increments. The diagram and relevant passage are reproduced above in
the "Editor's Introduction," Section 1.3.

CET. And therefore it is concluded, that the Fluxions of the Lines *AB*, *BC*, and *AC*, being in the last Ratio of their evanescent Increments, are proportional to the Sides of the Triangle *CET*, or, which is all one, of the Triangle *VBC* similar thereunto.* It is particularly remarked and insisted on by the great Author, that the Points *C* and *c* must not be distant one from another, by any the least Interval whatsoever: But that, in order to find the ultimate Proportions of the Lines *CE*, *Ec*, and *Cc* (*i.e.* the Proportions of the Fluxions or Velocities) expressed by the finite Sides of the Triangle *VBC*, the Points *C* and *c* must be accurately coincident, *i.e.* one and the same.[30] A Point therefore is considered as a Triangle, or a Triangle is supposed to be formed in a Point. Which to conceive seems quite impossible. Yet some there are, who, though they shrink at all other Mysteries, make no difficulty of their own, who strain at a Gnat and swallow a Camel.

§35. I know not whether it be worth while to observe, that possibly some Men may hope to operate by Symbols and Suppositions, in such sort as to avoid the use of Fluxions, Momentuums, and Infinitesimals after the following manner. Suppose x to be one Absciss of a Curve, and z another Absciss of the same Curve. Suppose also that the respective areas are x^3 and z^3: and that $z - x$ is the Increment of the Absciss, and $z^3 - x^3$ the Increment of the Area, without considering how great, or how small those Increments may be. Divide now $z^3 - x^3$ by $z - x$ and the Quotient will be $z^2 + zx + x^2$: and, supposing that z and x are equal,

---

* Introd. ad Quad. Curv.

[30] Berkeley's reference here is to Newton's declaration that "If the points *C* and *c* are at any small distance apart from each another, the straight line *CK* will be a small distance away from the tangent *CH*; in order that the line *CK* shall coincide with the tangent *CH* and so the last ratios of the lines *CE*, *Ec* and *Cc* be discovered, the points *C* and *c* must come together and entirely coincide. The most minute erors are not in mathematical matters to be scorned," (*Papers*, 8: 125). As in §§4 and 9 of *The Analyst*, Newton's insistence upon the accuracy of his methods gives Berkeley a significant rhetorical weapon.

this same Quotient will be $3x^2$ which in that case is the Ordinate, which therefore may be thus obtained independently of Fluxions and Infinitesimals. But herein is a direct Fallacy: for in the first place, it is supposed that the Abscisses z and x are unequal, without which supposition no one step could have been made; and in the second place, it is supposed they are equal; which is a manifest Inconsistency, and amounts to the same thing that hath been before considered.* And there is indeed reason to apprehend, that all Attempts for setting the abstruse and fine Geometry on a right Foundation, and avoiding the Doctrine of Velocities, Momentuums, &c. will be found impracticable, till such time as the Object and End of Geometry are better understood, than hitherto they seem to have been.[31] The great Author of the Method of Fluxions felt this Difficulty, and therefore he gave into those nice Abstractions and Geometrical Metaphysics, without which he saw nothing could be done on the received Principles; and what in the way of Demonstration he hath done with them the Reader will judge. It must, indeed, be acknowledged, that he used Fluxions, like the Scaffold of a building, as things to be laid aside or got rid of, as soon as finite Lines were found proportional to them. But then these finite Exponents are found by the help of Fluxions. Whatever therefore is got by such Exponents and Proportions is to be ascribed to Fluxions: which must therefore be previously understood. And what are these Fluxions? The Velocities of evanescent Increments? And what are these same evanescent Increments? They are neither finite Quantities, nor Quantities infinitely small, nor yet nothing. May we not call them the Ghosts of departed Quantities?

---

* *Sect.* 15.

[31] Queries 1 and 2 show that Berkeley conceives the object of geometry as the "proportion of assignable extensions," while the end of geometry is to "measure assignable finite extension."

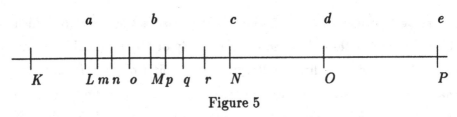

Figure 5

§36. Men too often impose on themselves and others, as if they conceived and understood things expressed by Signs, when in truth they have no Idea, save only of the very Signs themselves. And there are some grounds to apprehend that this may be the present Case. The Velocities of evanescent or nascent Quantities are supposed to be expressed, both by finite Lines of a determinate Magnitude, and by Algebraical Notes or Signs: but I suspect that many who, perhaps never having examined the matter, take it for granted, would upon a narrow scrutiny find it impossible, to frame any Idea or Notion whatsoever of those Velocities, exclusive of such finite Quantities and Signs. Suppose the line KP described by the Motion of a Point continually accelerated, and that in equal Particles of time the unequal Parts KL, LM, MN, NO, &c. are generated [Figure 5]. Suppose also that a, b, c, d, e, &c. denote the Velocities of the generating Point, at the several Periods of the Parts or Increments so generated. It is easy to observe that these Increments are each proportional to the sum of the Velocities with which it is described: That, consequently, the several sums of the Velocities, generated in equal Parts of Time, may be set forth by the respective Lines KL, LM, MN, &c. generated in the same times: It is likewise an easy matter to say, that the last Velocity generated in the first Particle of Time may be expressed by the Symbol a, the last in the second by b, the last generated in the third by c, and so on: that a is the Velocity of LM in *statu nascenti*, and b, c, d, e, &c. are the Velocities of the Increments MN, NO, OP, &c. in their respective nascent estates.

You may proceed, and consider these Velocities themselves as flowing or increasing Quantities, taking the Velocities of the Velocities, and the Velocities of the Velocities of the Velocities, *i.e.* the first, second, third, *&c.* Velocities *ad infinitum*: which succeeding Series of Velocities may be thus expressed,

$$a, (b - a), (c - 2b + a), (d - 3c + 3b - a), \&c.$$

which you may call by the names of first, second, third, fourth Fluxions.[32] And for an apter Expression you may denote the variable flowing Line KL, KM, KN, *&c.* by the Letter x; and the first Fluxions by $\dot{x}$, the second by $\ddot{x}$, the third by $\dddot{x}$, and so on *ad infinitum*.

§37. Nothing is easier than to assign Names, Signs, or Expressions to these Fluxions, and it is not difficult to compute and operate by means of such Signs. But it will be found much more difficult, to omit the Signs and yet retain in our Minds the things, which we suppose to be signified by them. To consider the Exponents, whether Geometrical, or Algebraical, or Fluxionary, is no difficult Matter, But to form a precise Idea of a third Velocity for instance, in it self and by it self, *Hoc opus, hic labor.* Nor indeed is it an easy point, to form a clear and distinct Idea of any Velocity at all, exclusive of and prescinding from all length of time and space; as also from all Notes, Signs or Symbols whatsoever. This, if I may be allowed to judge of others by my self, is impossible. To me it seems evident, that Measures and Signs are absolutely necessary, in order to conceive or reason about Velocities; and that, consequently,

---

[32] The construction which generates this sequence of differences is as follows: begin with a and obtain further terms by substituting the next velocity in the sequence $\{a, b, c, d, \ldots\}$ for its predecessor in the previous expression; then subtract the previous term from the resulting substitution to get the next term in the sequence of differences. The result is a recursively generated sequence of differences of successive velocities.

when we think to conceive the Velocities simply and in themselves, we are deluded by vain Abstractions.[33]

§38. It may perhaps be thought by some an easier Method of conceiving Fluxions, to suppose them the Velocities wherewith the infinitesimal Differences are generated. So that the first Fluxions shall be the Velocities of the first Differences, the second the Velocities of the second Differences, the third Fluxions the Velocities of the third Differences, and so on *ad infinitum*. But not to mention the insurmountable difficulty of admitting or conceiving Infinitesimals, and Infinitesimals of Infinitesimals, *&c.* it is evident that this notion of Fluxions would not consist with the great Author's view; who held that the minutest Quantity ought not to be neglected, that therefore the Doctrine of Infinitesimal Differences was not to be admitted in Geometry, and who plainly appears to have introduced the use of Velocities or Fluxions, on purpose to exclude or do without them.

§39. To others it may possibly seem, that we should form a juster Idea of Fluxions, by assuming the finite unequal isochronal Increments *KL*, *LM*, *MN*, *&c.* and considering them in *statu nascenti*, also their Increments in *statu nascenti*, and the nascent Increments of those Increments, and so on, supposing the first nascent Increments to be proportional to the first Fluxions or Velocities, the nascent Increments of those Increments to be proportional to the second Fluxions, the third nascent Increments to be proportional to the third Fluxions, and so onwards. And, as the first Fluxions are the Velocities of the first nascent Increments, so the second Fluxions may be conceived to be the Velocities of the second nascent Increments, rather than the Velocities of Velocities.

---

[33] This condemnation of "vain abstractions" recalls §§31-2 and returns in Query 29.

By which means the Analogy of Fluxions may seem better preserved, and the notion rendered more intelligible.

§40. And indeed it should seem, that in the way of obtaining the second or third Fluxion of an Equation, the given Fluxions were considered rather as Increments than Velocities. But the considering them sometimes in one Sense, sometimes in another, one while in themselves, another in their Exponents, seems to have occasioned no small share of that Confusion and Obscurity, which is found in the Doctrine of Fluxions. It may seem therefore, that the Notion might be still mended, and that instead of Fluxions of Fluxions, or Fluxions of Fluxions of Fluxions, and instead of second, third, or fourth, &c. Fluxions of a given Quantity, it might be more consistent and less liable to exception to say, the Fluxion of the first nascent Increment, *i.e.* the second Fluxion; the Fluxion of the second nascent Increment, *i.e.* the third Fluxion; the Fluxion of the third nascent Increment, *i.e.* the fourth Fluxion, which Fluxions are conceived respectively proportional, each to the nascent Principle of the Increment succeeding that whereof it is the Fluxion.

§41. For the more distinct Conception of all which it may be considered, that if the finite Increment LM^* be divided into the Isochronal Parts Lm, mn, no, oM; and the Increment MN into the Parts Mp, pq, qr, rN Isochronal to the former; as the whole Increments LM, MN are proportional to the Sums of their describing Velocities, even so the homologous Particles Lm, Mp are also proportional to the respective accelerated Velocities with which they are described. And as the Velocity with which Mp is generated, exceeds that with which Lm was generated, even so the Particle Mp exceeds the Particle Lm. And in general, as the Isochronal Velocities describing the Particles of MN exceed the

---

* *See the foregoing Scheme in Sect.* 36.

Isochronal Velocities describing the Particles of LM, even so the Particles of the former exceed the correspondent Particles of the latter. And
this will hold, be the said Particles ever so small. MN therefore will exceed LM if they are both taken in their nascent States: and that excess
will be proportional to the excess of the Velocity b above the Velocity
a. Hence we may see that this last account of Fluxions comes, in the
upshot, to the same thing with the first.[†]

§42. But notwithstanding what hath been said it must still be acknowledged, that the finite Particles Lm or Mp, though taken ever so
small, are not proportional to the Velocities a and b; but each to a Series
of Velocities changing every Moment, or which is the same thing, to an
accelerated Velocity, by which it is generated, during a certain minute
Particle of time: That the nascent beginnings or evanescent endings of finite Quantities, which are produced in Moments or infinitely small Parts
of Time, are alone proportional to Velocities: That, therefore, in order
to conceive the first Fluxions, we must conceive Time divided into Moments, Increments generated in those Moments, and Velocities proportional to those Increments: That in order to conceive second and third
Fluxions, we must suppose that the nascent Principles or momentaneous
Increments have themselves also other momentaneous Increments, which
are proportional to their respective generating Velocities: That the Velocities of these second momentaneous Increments are second Fluxions:
those of their nascent momentaneous Increments third Fluxions. And
so on *ad infinitum.*

§43. By subducting the Increment generated in the first Moment
from that generated in the second, we get the Increment of an Increment.
And by subducting the Velocity generating in the first Moment from
that generating in the second, we get the Fluxion of a Fluxion. In like

---

[†] *Sect.* 36.

manner, by subducting the Difference of the Velocities generating in the two first Moments from the excess of the Velocity in the third above that in the second Moment, we obtain the third Fluxion. And after the same Analogy we may proceed to fourth, fifth, sixth Fluxions, &c. And if we call the Velocities of the first, second, third, fourth Moments a, b, c, d, the Series of Fluxions will be as above,

$$a, \ (b-a), \ (c-2b+a), \ (d-3c+cb-a),$$

ad infinitum, i.e. $\dot{x}, \ddot{x}, \dddot{x}, \ddddot{x}$, *ad infinitum.*

§44. Thus Fluxions may be considered in sundry Lights and Shapes, which all seem equally difficult to conceive. And indeed, as it is impossible to conceive Velocity without time or space, without either finite length or finite Duration,* it must seem above the powers of men to comprehend even the first Fluxions. And if the first are incomprehensible, what shall we say of the second and third Fluxions, &c. ? He who can conceive the beginning of a beginning, or the end of an end, somewhat before the first or after the last, may be perhaps sharpsighted enough to conceive these things. But most Men will, I believe, find it impossible to understand them in any sense whatever.

§45. One would think that Men could not speak too exactly on so nice a Subject. And yet, as was before hinted, we may often observe that the Exponents of Fluxions or Notes representing Fluxions are confounded with the Fluxions themselves. Is not this the Case, when just after the Fluxions of flowing Quantities were said to be the Celerities of their increasing, and the second Fluxions to be the mutations of the first

---

* *Sect.* 31

Fluxions or Celerities, we are told that $\overset{..}{z}, \overset{.}{z}, z, \dot z, \ddot z, \overset{...}{z}$ † represents a Series of Quantities, whereof each subsequent Quantity is the Fluxion of the preceeding; and each foregoing is a fluent Quantity having the following one for its Fluxion?

§46. Divers Series of Quantities and Expressions, Geometrical and Algebraical, may be easily conceived, in Lines, in Surfaces, in Species, to be continued without end or limit. But it will not be found so easy to conceive a Series, either of mere Velocities or of mere nascent Increments, distinct therefrom and corresponding thereunto. Some perhaps may be led to think the Author intended a Series of Ordinates, wherein each Ordinate was the Fluxion of the preceeding and Fluent of the following, *i.e.* that the Fluxion of one Ordinate was itself the Ordinate of another Curve; and the Fluxion of this last Ordinate was the Ordinate of yet another Curve; and so on *ad infinitum.* But who can conceive how the fluxion (whether Velocity or nascent Increment) of an Ordinate should be it self an Ordinate? Or more than that each preceding Quantity or Fluent is related to its Subsequent or Fluxion, as the Area of a curvilinear Figure to its Ordinate; agreeably to what the Author remarks, that each preceding Quantity in such Series is as the Area of a curvilinear Figure, whereof the Absciss is z, and the Ordinate is the following Quantity?

§47. Upon the whole it appears that the Celerities are dismissed, and instead thereof Areas and Ordinates are introduced. But however

---

† De Quadratura Curvarum. [The passage Berkeley has in mind here reads as follows: "In the sequel I consider inteterminate quantities as increasing or decreasing by a perpetual motion, that is, as onwards or backwards flowing. I denote them by the letters z, y, x, v, and mark their fluxions, that is the speeds of increase, by the same letters with points on: $\dot z, \dot y, \dot x, \dot v$. . . . [s]o too these quantities can be considered as fluxions of others which I shall denote thus: $\ddot z \, \ddot y, \ddot x, \ddot v$, and these as fluxions of other ones $\overset{...}{z}, \overset{...}{y}, \overset{...}{x}, \overset{...}{v}$. . . So that. . . $\overset{...}{z}, \ddot z, \dot z, z, \dot z, \ddot z, \overset{...}{z}$. . . represent a series of quantities of which any latter one is the fluxion of that preceeding it, and any former the fluent quantity having the subsequent one as its fluxion," (*Papers*, 7: 509-11).]

expedient such Analogies or such Expressions may be found for facili-
tating the modern Quadratures, yet we shall not find any light given us
thereby into the original real nature of Fluxions; or that we are enabled
to frame from thence just Ideas of Fluxions considered in themselves.
In all this the general ultimate drift of the Author is very clear, but his
Principles are obscure. But perhaps those Theories of the great Author
are not minutely considered or canvassed by his Disciples; who seem
eager, as was before hinted, rather to operate than to know, rather to
apply his Rules and his Forms, than to understand his Principles and en-
ter into his Notions. It is nevertheless certain, that in order to follow him
in his Quadratures, they must find Fluents from Fluxions; and in order
to this, they must know how to find Fluxions from Fluents; and order to
find Fluxions, they must first know what Fluxions are. Otherwise they
proceed without Clearness and without Science. Thus the direct Method
preceeds the inverse, and the knowledge of the Principles is supposed in
both. But as for operating according to Rules, and by the help of general
Forms, whereof the original Principles and Reasons are not understood,
this is to be esteemed merely technical. Be the Principles therefore ever
so abstruse and metaphysical, they must be studied by whoever would
comprehend the Doctrine of Fluxions. Nor can any Geometrician have
a right to apply the Rules of the great Author, without first considering
his metaphysical Notions whence they were derived.[34] These how necessary soever in order to Science, which can never be attained without

---

[34] This mention of "abstruse and metaphysical" principles appears to be a
veiled reference to Newton's doctrines of absolute space and time, as set forth
in the famous Scholium to the Definitions in Book I of the *Principia*. Newton
gives little in the way of "metaphysical" discussion of the foundations of the
calculus, but his reliance on the kinematic conception of magnitudes and his
treatment of fluxions as velocities can be seen as resting upon his account of
space and time.

a precise, clear, and accurate Conception of the Principles, are never-theless by several carelessly passed over; while the Expressions alone are dwelt on and considered and treated with great Skill and Management, thence to obtain other Expressions by Methods, suspicious and indirect (to say the least) if considered in themselves, however recommended by Induction and Authority; two Motives which are acknowledged sufficient to beget a rational Faith and moral Persuasion, but nothing higher.

§48. You may possibly hope to evade the Force of all that hath been said, and to screen false Principles and inconsistent Reasonings, by a general Pretence that these Objections and Remarks are Meta-physical. But this is a vain Pretence. For the plain Sense and Truth of what is advanced in the foregoing Remarks, I appeal to the Under-standing of every unprejudiced intelligent Reader. To the same I appeal, whether the Points remarked upon are not the most incomprehensible Metaphysics. And Metaphysics not of mine, but your own. I would not be understood to infer, that your Notions are false or vain because they are Metaphysical. Nothing is either true or false for that Reason. Whether a Point be called Metaphysical or no avails little. The question is whether it be clear or obscure, right or wrong, well or ill deduced?

§49. Although momentaneous Increments, nascent and evanescent Quantities, Fluxions and Infinitesimals of all Degrees, are in truth such shadowy Entities, so difficult to imagine or conceive distinctly, that (to say the least) they cannot be admitted as Principles or Objects of clear and accurate Science: and although this obscurity and incomprehensibil-ity of your Metaphysics had been alone sufficient, to allay your Preten-sions to Evidence; yet it hath, if I mistake not, been further shewn, that your Inferences are no more just than your Conceptions are clear, and that your Logics are as exceptionable as your Metaphysics. It should seem therefore upon the whole, that your Conclusions are not attained by just Reasoning from clear Principles; consequently, that the Employment of modern Analysts, however useful in mathematical Calculations, and Constructions, doth not habituate and qualify the Mind to apprehend clearly and infer justly; and consequently, that you have no right in Virtue of such Habits, to dictate out of your proper Sphere, beyond which your Judgement is to pass for no more than that of other Men.

§50. Of a long time I have suspected, that these modern Analytics were not scientifical, and gave some Hints thereof to the public about twenty five years ago.[35] Since which time, I have been diverted by other Occupations, and imagined I might employ my self better than in deducing and laying together my Thoughts on so nice a Subject. And though of late I have been called upon to make good my Suggestions;[36] yet, as the Person who made this Call, doth not appear to think maturely enough to understand, either those Metaphysics which he would refute, or Mathematics which he would patronize, I should have spared my self the trouble of writing for his Conviction. Nor should I now have troubled you or my self with this Address, after so long an Intermission of these Studies; were it not to prevent, so far as I am able, your imposing on your self and others in Matters of much higher Moment and Concern.

---

[35] A reference to §§130-132 of the *Principles of Human Knowledge* (*Works*, 2: 101-102), where Berkeley suggests that the calculus stands in need of scrutiny and reform. He intimates that the doctrines of fluxions and infinitesimals can be done without, but promised that "this will be more clearly made out hereafter." A projected third part of the *Principles of Human Knowledge* dealing with mathematics and natural philosophy was never written, but we can assume that much of what Berkeley had in mind for that work appeared in *De Motu* and *The Analyst*.

[36] This reference is difficult to decipher. A. A. Luce suggests in a note that it is "Apparently a reference to Andrew Baxter's attack on Berkeley's philosophy in *An Enquiry into the nature of the human soul*, 1733," (*Works*, 4: 95). Baxter's comments on mathematics are, however, too brief to count as a serious challenge to Berkeley's account of the calculus. He does declare mathematics to be inconsistent with the Berkeleyan philosophy in §19 of the *Enquiry*, but only for the reason that "upon this scheme the object of [the] whole science is unphilosophically universal and abstract," (Baxter 1733, 307).

And, to the end that you may more clearly comprehend the Force and Design of the foregoing Remarks, and pursue them still further in your own Meditations, I shall subjoin the following Queries.

Query 1. Whether the Object of Geometry be not the Proportions of assignable Extensions? And whether, there be any need of considering Quantities either infinitely great or infinitely small?

Qu. 2. Whether the end of Geometry be not to measure assignable finite Extension? And whether this practical View did not first put Men on the study of Geometry?

Qu. 3. Whether the mistaking the Object and End of Geometry hath not created needless Difficulties, and wrong Pursuits in that Science?

Qu. 4. Whether Men may properly be said to proceed in a scientific Method, without clearly conceiving the Object they are conversant about, the End proposed, and the Method by which it is pursued?

Qu. 5. Whether it doth not suffice, that every assignable number of Parts may be contained in some assignable Magnitude? And whether it be not unnecessary, as well as absurd, to suppose that finite Extension is infinitely divisible?[37]

Qu. 6. Whether the Diagrams in a Geometrical Demonstration are not to be considered, as Signs, of all possible finite Figures, of all sensible and imaginable Extensions or Magnitudes of the same kind?

---

[37] This and the following Query recall Berkeley's critique of the doctrine of infinite divisibility in §§123-129 of the *Principles of Human Knowledge* (*Works*, 2: 97-101. There, he insists that no finite extension is infinitely divisible. Nevertheless, he concludes that because the lines and figures used in geometric demonstrations serve as signs for other lines and figures of different sizes, we can disregard the actual magnitude of any line or figure used in a demonstration. See Jesseph (1990) for more on Berkeley's philosophy of geometry. Similar points are raised in §17 and Queries 18 and 20.

Qu. 7. Whether it be possible to free Geometry from insuperable Difficulties and Absurdities, so long as either the abstract general Idea of Extension, or absolute external Extension be supposed its true Object?

Qu. 8. Whether the Notions of absolute Time, absolute Place, and absolute Motion be not most abstractedly Metaphysical? Whether it be possible for us to measure, compute, or know them?[38]

Qu. 9. Whether Mathematicians do not engage themselves in Disputes and Paradoxes, concerning what they neither do nor can conceive? And whether the Doctrine of Forces be not a sufficient Proof of this?[*]

Qu. 10. Whether in Geometry it may not suffice to consider assignable finite Magnitude, without concerning our selves with Infinity? And whether it would not be righter to measure large Polygons having finite Sides, instead of Curves, than to suppose Curves are Polygons of infinitesimal Sides, a Supposition neither true nor conceivable?[39]

Qu. 11. Whether many Points which are not readily assented to, are not nevertheless true? And whether those in the two following Queries may not be of that Number?

Qu. 12. Whether it be possible, that we should have had an Idea or Notion of Extension prior to Motion? Or whether if a Man had never perceived Motion, he would ever have known or conceived one thing to be distant from another?

---

[38] *Cf.* Berkeley's remarks on absolute space and time in §§110-117 of the *Principles of Human Knowledge* (*Works*, 2: 89-94) and §§52-66 of *De Motu*.

[*] *See a* Latin *treatise* De Motu, *published at* London, *in the year* 1721.

[39] Cf. §5 for the reference to L'Hôpital's doctrine that curves are polygons with infinitesimal sides. The complaint voiced in this Query appears in entry 527 of the *Philosophical Commentaries*: "W^t do the Mathematicians mean by Considering Curves as Polygons? either they are Polygons or are they are not. If the are why do they give them the Name of Curves? why do they not constantly call them Polygons & treat them as such. If they are not polygons I think it absurd to use polygons in their Stead. W^t. is this but to pervert language and adapt and idea to a name that belongs not to it but to a different idea?" *Works*, 1: 65-6.

Qu. 13. Whether Geometrical Quantity hath coexistent Parts? And whether all Quantity be not in a flux as well as Time and Motion?

Qu. 14. Whether Extension can be supposed an Attribute of a Being immutable and eternal?[40]

Qu. 15. Whether to decline examining the Principles, and unravelling the Methods used in Mathematics, would not shew a bigotry in Mathematicians?

Qu. 16. Whether certain Maxims do not pass current among Analysts, which are shocking to good Sense? And whether the common Assumption that a finite Quantity divided by nothing is infinite be not of this Number?[41]

Qu. 17. Whether the considering Geometrical Diagrams absolutely or in themselves, rather than as Representatives of all assignable Magnitudes or Figures of the same kind, be not a principal Cause of the

---

[40] Newton's doctrine of absolute space poses theological problems which Berkeley here adumbrates. If absolute space is eternal, uncreated, and immutable then it must be either an attribute of God or something semi-divine in itself. Newton suggests that absolute space and time are aspects of the Deity in the "General Scholium" to the *Principia*: "Since every particle of space is *always* and every indivisible moment of duration *everywhere*, certainly the Maker and Lord of all things cannot be *never* and *nowhere*," (*Principia*, 2: 545). Berkeley makes similar points in entry 298 of the *Philosophical Commentaries* (*Works*, 1: 37), §117 of the *Principles of Human Knowledge* (*Works*, 2: 93-4), and §56 of *De Motu*.

[41] This query recalls Berkeley's discussion of Wallis's methods in his early essay "Of Infinites." There, he insists that when Wallis treats the quotient $\frac{1}{0}$ as an infinite magnitude, he is likewise committed to the doctrine that infinitesimals are nothing: "Since, therefore, unity, *i.e.* any finite line divided by 0, gives the asymptote of an hyperbola, *i.e.* a line infinitely long, it necessarily follows that a finite line divided by an infinite line gives 0 in the quotient, *i.e.* that the *pars infinitesima* of a finite line is just nothing. For by the nature of division the dividend divided by the quotient gives the divisor," (*Works*, 4: 236). A similar line of argument is pursued in Query 40.

supposing finite Extension infinitely divisible; and of all the Difficulties and Absurdities consequent thereupon?[42]

Qu. 18. Whether from Geometrical Propositions being general, and the Lines in Diagrams being therefore general Substitutes or Representatives, it doth not follow that we may not limit or consider the number of Parts, into which such particular Lines are divisible?

Qu. 19. When it is said or implied, that such a certain Line delineated on Paper contains more than any assignable number of Parts, whether any more in truth ought to be understood, than that it is a Sign indifferently representing all finite Lines, be they ever so great. In which relative Capacity it contains, *i.e.* stands for more than any assignable number of Parts? And whether it be not altogether absurd to suppose a finite line, considered in it self or in its own positive Nature, should contain an infinite number of Parts?

Qu. 20. Whether all Arguments for the infinite Divisibility of finite Extension do not suppose and imply, either general abstract Ideas or absolute external Extension to be the Object of Geometry? And, therefore, whether, along with those Suppositions, such Arguments also do not cease and vanish?

Qu. 21. Whether the supposed infinite Divisibility of finite Extension hath not been a Snare to Mathematicians, and a Thorn in their Sides? And whether a Quantity infinitely diminished and a Quantity infinitely small are not the same thing?

Qu. 22. Whether it be necessary to consider Velocities of nascent or evanescent Quantities, or Moments, or Infinitesimals? And whether the introducing of Things so inconceivable be not a reproach to Mathematics?

---

[42] For this and the following four Queries, *cf.* §§123-129 of the *Principles of Human Knowledge* (*Works*, 2: 97-101), where Berkeley considers the nature of geometry.

Qu. 23. Whether Inconsistencies can be Truths? Whether Points repugnant and absurd are to be admitted upon any Subject, or in any Science? And whether the use of Infinities ought to be allowed, as a sufficient Pretext and Apology, for the admitting of such Points in Geometry?

Qu. 24. Whether a Quantity be not properly said to be known, when we know its Proportion to given Quantities? And whether this Proportion can be known, but by Expressions or Exponents, either Geometrical, Algebraical, or Arithmetical? And whether Expressions in Lines or Species can be useful but so far forth as they are reducible to Numbers?

Qu. 25. Whether the finding out proper Expressions or Notations of Quantity be not the most general Character and Tendency of the Mathematics? And Arithmetical Operation that which limits and defines their Use?

Qu. 26. Whether Mathematicians have sufficiently considered the Analogy and Use of Signs? And how far the specific limited Nature of things corresponds thereto?

Qu. 27. Whether because, in stating a general Case of pure Algebra, we are at full liberty to make a Character denote, either a positive or a negative Quantity, or nothing at all, we may therefore in a geometrical Case, limited by Hypotheses and Reasonings from particular Properties and Relations of Figures, claim the same License?[43]

---

[43] Again, Berkeley's makes a distinction between algebra and the calculus. Algebraic reasoning employs arbitrary characters which may or may not stand for specific, known quantities. Geometric reasoning, however, concerns the relations between finite magnitudes, so the application of algebra to geometry must respect the "hypotheses and reasonings" pertaining to figures. Cf. Queries 41, 45, and 46 for more on Berkeley's conception of algebra and its relationship to geometry and the calculus.

Qu. 28. Whether the Shifting of the Hypothesis, or (as we may call it) the *fallacia Suppositionis* be not a Sophism, that far and wide infects the modern Reasonings, both in the mechanical Philosophy and in the abstruse and fine Geometry?

Qu. 29. Whether we can form an Idea or Notion of Velocity distinct from and exclusive of its Measures, as we can of Heat distinct from and exclusive of the Degrees on the Thermometer, by which it is measured? And whether this be not supposed in the Reasonings of modern Analysts?

Qu. 30. Whether Motion can be conceived in a Point of Space? And if Motion cannot, whether Velocity can? And if not, whether a first or last Velocity can be conceived in a mere Limit, either initial or final, of the described Space?

Qu. 31. Where there are no Increments, whether there can be any *Ratio* of Increments? Whether Nothings can be considered as proportional to real Quantities? Or whether to talk of their Proportions be not to talk Nonsense? Also in what Sense we are to understand the Proportion of a Surface to a Line, of an Area to an Ordinate? And whether Species or Numbers, though properly expressing Quantities which are not homogeneous, may yet be said to express their Proportion to each other?[44]

Qu. 32. Whether if all assignable Circles may be squared, the Circle is not, to all intents and purposes, squared as well as the Parabola? Or whether a parabolical Area can in fact be measured more accurately than a Circular?

---

[44] The emphasis here on the homogeneity of magnitudes in a ratio states the key requirement in the classical theory of ratios: two quantities can form a ratio if and only if successive (finite) multiplication of one can make it exceed the other. There is no ratio of a line to an area, but the procedures of the infinitesimal calculus violate these restrictions.

Qu. 33. Whether it would not be righter to approximate fairly, than to endeavour at Accuracy by Sophisms?

Qu. 34. Whether it would not be more decent to proceed by Trials and Inductions, than to pretend to demonstrate by false Principles?

Qu. 35. Whether there be not a way of arriving at Truth, although the Principles are not scientific, nor the Reasoning just? And whether such a way ought to be called a Knack or a Science?

Qu. 36. Whether there can be Science of the Conclusion, where there is not Evidence of the Principles? And whether a Man can have Evidence of the Principles, without understanding them? And therefore whether the Mathematicians of the present Age act like men of Science, in taking so much more pains to apply their Principles, than to understand them?[45]

Qu. 37. Whether the greatest Genius wrestling with false Principles may not be foiled? And whether accurate Quadratures can be obtained without new *Postulata* or Assumptions? And if not, whether those which are intelligible and consistent ought not to be preferred to the contrary? *See Sect.* 28 *and* 29.

Qu. 38. Whether tedious Calculations in Algebra and Fluxions be the likeliest Method to improve the Mind? And whether Mens being accustomed to reason altogether about Mathematical Signs and Figures, doth not make them at a loss how to reason without them?

Qu. 39. Whether, whatever readiness Analysts acquire in stating a Problem, or finding apt Expressions for Mathematical Quantities, the

---

[45] The first London edition prints the either incomplete ' ence' or the word 'Science' for the two occurences of the word the word 'Evidence' in this query. The printing is then corrected to 'Evidence' in the 1734 Dublin edition, which is consistent with a hand-corrected version of the first London edition known to have been made by Berkeley himself. Strangely, the posthumous second London edition repeats the error of printing 'Science' for 'Evidence' in this query. For more on this, see Keynes (1976, 65-71).

same doth necessarily infer a proportionable ability in conceiving and expressing other Matters?

Qu. 40. Whether it be not a general Case or Rule, that one and the same Coefficient dividing equal Products gives equal Quotients? And yet whether such Coefficient can be interpreted by *o* or nothing? Or whether any one will say, that if the Equation $2 \times o = 5 \times o$, be divided by *o*, the Quotients on both Sides are equal? Whether therefore a Case may not be general with respect to all Quantities and yet not extend to Nothings, or include the Case of Nothing? And whether the bringing Nothing under the Notion of Quantity may not have betrayed Men into false Reasoning?

Qu. 41. Whether in the most general Reasonings about Equalities and Proportions, Men may not demonstrate as well as in Geometry? Whether in such Demonstrations, they are not obliged to the same strict Reasonings as in Geometry? And whether such their Reasonings are not deduced from the same Axioms with those in Geometry? Whether therefore Algebra be not as truly a Science as Geometry?

Qu. 42. Whether Men may not reason in Species as well as in Words? Whether the same Rules of Logic do not obtain in both Cases? And whether we have not a right to expect and demand the same Evidence in both?

Qu. 43. Whether an Algebraist, Fluxionist, Geometrician or Demonstrator of any kind can expect indulgence for obscure Principles or incorrect Reasonings? And whether an Algebraical Note or Species can at the end of a Process be interpreted in a Sense, which could not have been substituted for it at the beginning? Or whether any particular Supposition can come under a general Case which doth not consist with the reasoning thereof?

Qu. 44. Whether the Difference between a mere Computer and a Man of Science be not, that the one computes on Principles clearly conceived, and by Rules evidently demonstrated, whereas the other doth not?

Qu. 45. Whether, although Geometry be a Science, and Algebra allowed to be a Science, and the Analytical a most excellent Method, in the Application nevertheless of the Analysis to Geometry, Men may not have admitted false Principles and wrong Methods of Reasoning?

Qu. 46. Whether, although Algebraical Reasonings are admitted to be ever so just, when confined to Signs or Species as general Representatives of Quantity, you may not nevertheless fall into Error, if, when you limit them to stand for particular things, you do not limit your self to reason consistently with the Nature of such particular things? And whether such Error ought to be imputed to pure Algebra?

Qu. 47. Whether the View of modern Mathematicians doth not rather seem to be the coming at an Expression by Artifice, than at coming at Science by Demonstration?

Qu. 48. Whether there may not be sound Metaphysics as well as unsound? Sound as well as unsound Logic? And whether the modern Analytics may not be brought under one of these Denominations, and which?

Qu. 49. Whether there be not really a *Philosophia prima*, a certain transcendental Science superior to and more extensive than Mathematics, which it might behove our modern Analysts rather to learn than despise?

Qu. 50. Whether ever since the recovery of Mathematical Learning, there have not been perpetual Disputes and Controversies among the

Mathematicians? And whether this doth not disparage the Evidence of their Methods?[46]

Qu. 51. Whether anything but Metaphysics and Logic can open the Eyes of Mathematicians and extricate them out of their Difficulties?

Qu. 52. Whether upon the received Principles a Quantity can by any Division or Subdivision, though carried ever so far, be reduced to nothing?[47]

Qu. 53. Whether, if the end of Geometry be Practice, and this Practice be Measuring, and we measure only assignable Extensions, it will not follow that unlimited Approximations compleatly answer the Intention of Geometry?

Qu. 54. Whether the same things which are now done by Infinites may not be done by finite Quantities? And whether this would not be a great Relief to the Imaginations and Understandings of Mathematical Men?

Qu. 55. Whether those Philomathematical Physicians, Anatomists, and Dealers in the animal Oeconomy, who admit the Doctrine of Fluxions with an implicit Faith, can with a good grace insult other Men for believing what they do not comprehend?

Qu. 56. Whether the Corpuscularian, Experimental, and Mathematical Philosophy so much cultivated in the last Age, hath not too

---

[46] The controversies Berkeley refers to here are several and presumably include: Christopher Clavius (1537-1612) and Jacques Peletier (1517-158) on the "angle of contact" between circle and tangent, Hobbes and Wallis on the quadrature of the circle, Wallis and Barrow on the nature of proportions and the relationship between arithmetic and geometry, Cavalieri and Paul Guldin (1577-1643) on the method of indivisibles, and finally Leibniz and Bernard Nieuwentijt (1654-1718) on higher-order differentials. He catalogues such controversies in §15 of Dialogue VII of the *Alciphron* when he speaks of "difficulties and disputes" which "have sprung up in geometry about the nature of the angle of contact, the doctrine of proportions, of indivisibles, infinitesimals, and divers other points," (*Works*, 3: 308).

[47] *Cf.* Query 5 and §17.

much engrossed Mens Attention; some part whereof it might have usefully employed?

Qu. 57. Whether from this, and other concurring Causes, the Minds of speculative Men have not been born downward, to the debasing and stupifying of the higher Faculties? And whether we may not hence account for that prevailing Narrowness and Bigotry among many who pass for Men of Science, their Incapacity for things Moral, Intellectual, or Theological, their Proneness to measure all Truths by Sense and Experience of animal Life?

Qu. 58. Whether it be really an Effect of Thinking, that the same Men admire the great Author for his Fluxions, and deride him for his Religion?[48]

Qu. 59. If certain Philosophical Virtuosi of the present Age have no Religion, whether it can be said to be for want of Faith?

Qu. 60. Whether it be not a juster way of reasoning, to recommend Points of Faith from their Effects, than to demonstrate Mathematical Principles by their Conclusions?

Qu. 61. Whether it be not less exceptionable to admit Points above Reason than contrary to Reason?

Qu. 62. Whether Mysteries may not with better right be allowed of in Divine Faith than in Humane Science?

Qu. 63. Whether such Mathematicians as cry out against Mysteries, have ever examined their own Principles?

---

[48] Newton's theological interests were well-known and would presumably have been something of an embarassment to a free-thinking analyst who professed admiration for Newton's calculus of fluxions. His 1728 *Chronology of Ancient Kingdoms Ammended* is mentioned in §22 of Dialogue VI of the *Alciphon* and dismissed by the free-thinker Alciphron when he remarks that "It hath been observed by ingenious men that Sir Isaac Newton," though a layman, was deeply prejudiced: witness his great regard to the Bible," (*Works,* 3: 264).

Qu. 64. Whether Mathematicians, who are so delicate in religious Points, are strictly scrupulous in their own Science? Whether they do not submit to Authority, take things upon Trust, believe Points inconceivable? Whether they have not their Mysteries, and what is more, their Repugnancies and Contradictions?

Qu. 65. Whether it might not become Men, who are puzzled and perplexed about their own Principles, to judge warily, candidly, and modestly concerning other Matters?

Qu. 66. Whether the modern Analytics do not furnish a strong *argumentum ad hominem*, against the Philomathematical Infidels of these Times?

Qu. 67. Whether it follows from the abovementioned Remarks, that accurate and just Reasoning is the peculiar Character of the present Age? And whether the modern Growth of Infidelity can be ascribed to a Distinction so truly valuable?

F I N I S

Index to *De Motu*

Index to *The Analyst*

Synthese Historical Library
Texts and Studies in the History of Logic and Philosophy

Series Editor: Norman Kretzmann (*Cornell University*)

1. M.T. Beonio-Brocchieri Fumagalli: *The Logic of Abelard.* Translated from Italian by S. Pleasance. 1969   ISBN 90-277-0068-0
2. G. W. Leibniz: *Philosophical Papers and Letters.* A Selection, translated and edited, with an Introduction, by L. E. Loemker. 2nd ed., 2nd printing. 1976
     ISBN 90-277-0008-8
3. E. Mally: *Logische Schriften.* Grosses Logikfragment – Grundgesetze des Sollens. Herausgegeben von K. Wolf und P. Weingartner. 1971   ISBN 90-277-0174-1
4. L. W. Beck (ed.): *Proceedings of the Third International Kant Congress.* 1972
     ISBN 90-277-0188-1
5. B. Bolzano: *Theory of Science.* A Selection with an Introduction by J. Berg. Translated from German by B. Terrell. 1973   ISBN 90-277-0248-9
6. J. M. E. Moravcsik (ed.): *Patterns in Plato's Thought.* 1973   ISBN 90-277-0286-1
7. Avicenna: *The Propositional Logic.* A Translation from *Al-Shifā': al-Qiyās*, with Introduction, Commentary and Glossary by N. Shehaby. 1973   ISBN 90-277-0360-4
8. D. P. Henry: *Commentary on* De Grammatico. *The Historical-Logical Dimensions of a Dialogue of St. Anselms's.* 1974   ISBN 90-277-0382-5
9. J. Corcoran (ed.): *Ancient Logic and its Modern Interpretations.* 1974
     ISBN 90-277-0395-7
10. E. M. Barth: *The Logic of the Articles in Traditional Philosophy.* A Contribution to the Study of Conceptual Structures. 1974   ISBN 90-277-0350-7
11. J. Hintikka: *Knowledge and the Known.* Historical Perspectives in Epistemology. 1974
     ISBN 90-277-0455-4
12. E. J. Ashworth: *Language and Logic in the Post-Medieval Period.* 1974
     ISBN 90-277-0464-3
13. Aristotle: *The Nicomachean Ethics.* Translation with Commentaries and Glossary by H. G. Apostle. 1974   ISBN 90-277-0569-0
14. R. M. Dancy: *Sense and Contradiction.* A Study in Aristotle. 1975
     ISBN 90-277-0565-8
15. W. R. Knorr: *The Evolution of the Euclidean Elements.* A Study of the Theory of Incommensurable Magnitudes and its Significance for Early Greek Geometry. 1975
     ISBN 90-277-0509-7
16. Augustine: *De Dialectica.* Translated with Introduction and Notes by B. D. Jackson from the Text newly edited by J. Pinborg. 1975   ISBN 90-277-0538-9
17. Á. Szabó: *The Beginnings of Greek Mathematics.* Translated from German. 1978
     ISBN 90-277-0819-3
18. Juan Luis Vives: *Against the Pseudodialecticians.* A Humanist Attack on Medieval Logic. Texts (in Latin), with Translation, Introduction and Notes by R. Guerlac. 1979
     ISBN 90-277-0900-9
19. Peter of Ailly: *Concepts and Insolubles.* An Annotated Translation (from Latin) by P. V. Spade. 1980   ISBN 90-277-1079-1
20. S. Knuuttila (ed.): *Reforging the Great Chain of Being.* Studies of the History of Modal Theories. 1981   ISBN 90-277-1125-9
21. J. V. Buroker: *Space and Incongruence.* The Origin of Kant's Idealism. 1981
     ISBN 90-277-1203-4

Synthese Historical Library

22. Marsilius of Inghen: *Treatises on the Properties of Terms.* A First Critical Edition of the *Suppositiones, Ampliationes, Appellationes, Restrictiones* and *Alienationes* with Introduction, Translation, Notes and Appendices by E. P. Bos. 1983
ISBN 90-277-1343-X
23. W. R. de Jong: *The Semantics of John Stuart Mill.* 1982   ISBN 90-277-1408-8
24. René Descartes: *Principles of Philosophy.* Translation with Explanatory Notes by V. R. Miller and R. P. Miller. 1983   ISBN 90-277-1451-7
25. T. Rudavsky (ed.): *Divine Omniscience and Omnipotence in Medieval Philosophy.* Islamic, Jewish and Christian Perspectives. 1985   ISBN 90-277-1750-8
26. William Heytesbury: *On Maxima and Minima.* Chapter V of *Rules for Solving Sophismata*, with an Anonymous 14th-century Discussion. Translation from Latin with an Introduction and Study by J. Longeway. 1984   ISBN 90-277-1868-7
27. Jean Buridan's *Logic. The Treatise on Supposition. The Treatise on Consequences.* Translation from Latin with a Philosophical Introduction by P. King. 1985
ISBN 90-277-1918-7
28. S. Knuuttila and J. Hintikka (eds.): *The Logic of Being.* Historical Studies. 1986
ISBN 90-277-2019-3
29. E. Sosa (ed.): *Essays on the Philosophy of George Berkeley.* 1987
ISBN 90-277-2405-9
30. B. Brundell: *Pierre Gassendi: From Aristotelianism to a New Natural Philosophy.* 1987   ISBN 90-277-2428-8
31. Adam de Wodeham: *Tractatus de indivisibilibus.* A Critical Edition with Introduction, Translation, and Textual Notes by R. Wood. 1988   ISBN 90-277-2424-5
32. N. Kretzmann (ed.): *Meaning and Inference in Medieval Philosophy.* Studies in Memory of J. Pinborg (1937–1982). 1988   ISBN 90-277-2577-2
33. S. Knuuttila (ed.): *Modern Modalities.* Studies of the History of Modal Theories from Medieval Nominalism to Logical Positivism. 1988   ISBN 90-277-2678-7
34. G. F. Scarre: *Logic and Reality in the Philosophy of John Stuart Mill.* 1988
ISBN 90-277-2739-2
35. J. van Rijen: *Aspects of Aristotle's Logic of Modalities.* 1989   ISBN 0-7923-0048-3
36. L. Baudry: *The Quarrel over Future Contingents (Louvain 1465–1475).* Unpublished Latin Texts collected and translated in French by L. Baudry. Translated from French by R. Guerlac. 1989   ISBN 0-7923-0454-5

THE NEW SYNTHESE HISTORICAL LIBRARY
Texts and Studies in the History of Philosophy

37. S. Payne: *John of the Cross and the Cognitive Value of Mysticism.* An Analysis of Sanjuanist Teaching and its Philosophical Implications for Contemporary Discussions of Mystical Experience. 1990   ISBN 0-7923-0707-0
38. D.D. Merrill: *Augustus De Morgan and the Logic of Relations.* 1990
ISBN 0-7923-0758-5
39. H. T. Goldstein (ed.): *Averroes' Questions in Physics.* 1991   ISBN 0-7923-0997-9
40. George Berkeley: De Motu *and* The Analyst. A Modern Edition with Introductions and Commentary by Douglas M. Jesseph. 1992   ISBN 0-7923-1520-0

Kluwer Academic Publishers – Dordrecht / Boston / London